Russia and the Changing Character of Conflict

Russia and the Changing Character of Conflict

Tracey German

Rapid Communications in Conflict and Security Series
General Editor: Thomas G. Mahnken
Founding Editor: Geoffrey R.H. Burn

CAMBRIA
PRESS

Amherst, New York

Library of Congress Cataloging-in-Publication Data

Names: German, Tracey C., 1971- author.

Title: Russia and the changing character of conflict / Tracey German.

Description: Amherst : Cambria Press, [2023] |
Series: Rapid communications in conflict and security series |
Includes bibliographical references and index. |
Summary: "Russia's actions in and around Ukraine in 2014, as well as its activities in Syria
and further afield, sparked renewed debate about the character of war and armed conflict,
and whether it was undergoing a fundamental shift. Since 2014 there has been wide-ranging
discussion about Russia's "new way of war", with labels such as hybrid warfare, grey-
zone operations and the Gerasimov doctrine dominating Western analyses. However, there
has been scant analysis of Russian perspectives on the changing character of conflict and
what future wars may look like: Western attempts to understand how and why Russia
uses force have tended to rely upon mirror-imaging and an expectation of similar strategic
behaviors. This book explores Russian views of the changing character of conflict and the
debates that have emerged about how future wars might evolve. It sets out the trends and
debates in Russian military thought, outlining the implications of Russian conclusions
regarding the characteristics of contemporary and future conflict"-- Provided by publisher.

Identifiers: LCCN 2022058751 (print) | LCCN 2022058752 (ebook) |
ISBN 9781621966739 (library binding) | ISBN 9781621966753 (paperback)
ISBN 9781621966845 (pdf) | ISBN 9781621966852 (epub)

Subjects: LCSH: Russia (Federation)--Military policy--21st century. | Russia (Federation)--
Foreign relations--21st century. | Russia (Federation)--History, Military--21st century.
| Hybrid warfare--Russia (Federation)--History--21st century. | Ukraine Conflict, 2014-

Classification: LCC UA770 .G37 2023 (print) | LCC UA770 (ebook) |
DDC 355/.033547--dc23/eng/20230206

LC record available at https://lccn.loc.gov/2022058751

LC ebook record available at https://lccn.loc.gov/2022058752

TABLE OF CONTENTS

List of Tables... vii

Introduction.. 1

Part I: Lessons Learned.. 13

Chapter 1: The Evolution of Military Thought......................... 15

Chapter 2: Observing Western Interventions........................... 53

Chapter 3: Operational Experience in the Post-Soviet Era............. 81

Part II: Continuity And Change....................................... 119

Chapter 4: High-Tech Futures.. 121

Chapter 5: Undermining the Will to Resist............................ 155

Chapter 6: All Available Means?....................................... 191

Chapter 7: Learning the Wrong Lessons?................................ 217

Conclusion.. 235

Glossary.. 247

Index... 251

Cambria Rapid Communications in Conflict and Security
      Series.. 267

List of Tables

Table 1: Key contributors to Russian military science
debates 30

Table 2: Slipchenko's Generations of Warfare 60

Russia and the Changing Character of Conflict

Introduction

Russia's actions in and around Ukraine in 2014, as well as its activities in Syria and farther afield, sparked renewed debate about the character of war and armed conflict and whether it was undergoing a fundamental shift. Since 2014 there has been wide-ranging discussion about Russia's "new way of war," with labels such as hybrid warfare, grey-zone operations, and the Gerasimov Doctrine dominating Western analyses. However, there has been scant analysis of Russian perspectives on the changing character of conflict and what future wars may look like. Western attempts to understand how and why Russia uses force have tended to rely upon mirror-imaging and an expectation of similar strategic behaviors. This book explores Russian views of the changing character of conflict and the debates that have emerged about how future wars might evolve. It seeks to encourage a greater understanding of Russian military thought, the range of perspectives a peer competitor holds, and the particular analytical processes that take place. It sets out the trends and debates in Russian military thought, outlining the implications of Russian conclusions regarding the characteristics of contemporary and future conflict. One of the enduring features of conflict over the centuries has been its state of flux. This perpetual state of evolution requires states to regularly monitor how military force is being wielded, either by allies or adversaries, in order to be able to plan and prepare for future war. The experiences of individual states foster different visions of future conflict and how states envisage military force being used, either by themselves or potential adversaries. It is vital to understand the process of observation and assessment that other states are engaged in. For states such as Russia, the lessons from the Western interventions of the twenty-first century have been instructive, shaping its perceptions of the changing character of conflict and the implications for its military.

It is important to differentiate between the character and nature of war and conflict. The nature of war refers to its enduring essence, what differentiates it from other activities. War's nature is unchanging: it is a violent human activity undertaken for political purpose; Carl von Clausewitz's act of violence intended to compel an adversary to submit to one's will. In contrast, the character of war and conflict continually evolves, reflecting the specific societal, technological, political, and historical context. Thus, warfare, and the means by which war is fought, is also continually changing, influenced by factors such as technology and operational innovation. The character of conflict is a principal topic of debate among Russian military theorists, and there is a long tradition of rigorous military strategic debate in Russia, as well as an emphasis on the systematic, scientific study of the theoretical foundations of war and conflict. The issue of foresight and forecasting the future character of war is an enduring concern, reflected in the writings of serving or retired military officers in open-source publications, many of which focus on the characteristics of conflict and the security threats facing Russia, particularly those believed to derive from the US. Common themes in these publications include a focus on the lessons that Russia could draw from Western interventions in places such as Serbia, Afghanistan, Iraq, and Libya, as well as Russia's own military experiences since 1991, and the perceived threat to Russian national security from internal instability, subversion, and regime change initiated by external actors, including "color revolutions" (*tsvetnaya revolyutsiya*[1]).

The belief that Russia has been constantly threatened throughout its history constitutes an enduring element of the country's military thought, shaping its worldview and self-image. Throughout history, Russian strategic thinking and threat perceptions have been shaped by a persistent fear of external aggression, hostile encirclement, and deliberate interference in the country's internal affairs. The worldview of the political and security elites is also shaped by the sheer size of territory, which has endured a number of invasions over the centuries, from the thirteenth-century Mongol invasion to that of Napoleon's Grand Armée at the beginning of the nineteenth century and the attack from Germany in June 1941, all of which prompt a sense of strategic vulnerability and fear of being taking by surprise. This sense of vulnerability has been exacerbated by an enduring preoccupation with technological inferiority compared to adversaries.[2] There was a prevailing anxiety evident in both Russian and Soviet strategic thought that enemies would take advantage of any weakness, creating an expectation of surprise attacks and a fear of encirclement and deception. This contributes to the notion of Russia as a "besieged fortress" that is surrounded by enemies and needs to be prepared for an attack at any time. The German invasion of 1941 is frequently cited as an example that should never be repeated. Russian military science, particularly its emphasis on foresight and forecasting the future character of war, is a deliberate countermeasure to this fear of surprise. Reflecting this pattern, Igor M. Popov and Musa M. Khamzatov contend that the military needs to be prepared for any eventuality, writing that the Russian leadership needs to be prepared for "multiple scenarios...and for any military conflict with a variety of adversaries in a range of conditions."[3] They reason that it is naive to believe that there will always be intelligence warnings of an opponent's preparations for war, arguing that the last time Russia believed this was in 1941, when Germany invaded. Russia's history and experience of unexpected regime collapses (the Russian Revolution in 1917, the fall of the Berlin Wall and collapse of the outer Soviet empire in 1989, and the collapse of the USSR in 1991) has fostered a belief that strategic decision makers should expect the unexpected.

As a result, there is considerable effort devoted to the systematic analysis of war and conflict, with a particular focus on the character of conflict, as well as the forms and methods that are used in them, in order to mitigate the element of surprise. Uncertainty and risk have always been an integral part of conflict; however, the transformation of contemporary conflict has increased their prevalence and efficacy. The increasing complexity of the global security environment in the twenty-first century has heightened the tension between the unpredictability of future war and the need for predictability to aid planning at the state level. Attempts to predict the character of future war, what it may look like, and the means that an adversary may use are an enduring concern of Russian military theorists; thus, there is a significant focus on analyzing trends in military developments across the world and what they may mean for Russia. During the Cold War, Soviet military thought focused on the impact of technological change on the balance of power between the USSR and the US, initially concentrating on the central role of nuclear weapons. A collection of articles published in the 1960s with the title *Problems of the Revolution in Military Affairs* analyzed the impact of nuclear weapons on warfare.[4] Also in the 1960s, Vasily Sokolovsky explored the forms and methods of warfare and the impact of new technologies on traditional components of military art.[5] Under the leadership of the Chief of the General Staff (CGS) Marshal Nikolai Ogarkov, Soviet military theorists in the 1980s identified a "military-technical revolution" (*voenno-technicheskaya revolyutsiya*), a fundamental shift in the character and conduct of military operations driven by advanced nuclear weapons, the development of long-range precision strike conventional weapons, and information technology.[6] Used to argue that technology on its own is insufficient to drive major military change, the term "military-technical revolution" was a precursor to the idea of a "revolution in military affairs" (RMA), which gained prominence in the US in the 1990s. Military theorists argued that RMAs require operational innovation and changes in doctrine and organization, as well as new technology. The end of the Cold War and the 1991 Gulf War renewed the debate about RMAs and the impact of new technologies on the conduct of war, with disagreement about both definition and occurrence.[7]

Russian military thought continues to be infused with an enduring belief that technology remains a key determinant of how war is fought, as evidenced by the focus on network-centric warfare (NCW, *setetsentrich-eskaya voina*), precision-strike, and information technologies. Russian experts believe that there has been a change in the forms and methods (*formyi i sposobyi*) that actors use to wage war, with developments in both the military and nonmilitary realms. This reflects a prevalent view among Russian military theorists in the early twenty-first century that the character of conflict is undergoing a process of dynamic change. During the mid-to-late 2000s there was a consistent number of articles in Russian military publications, such as *Voennaya Mysl'* and *Voenno-promyishlenniyi kur'er* (*VPK*), focusing on the characteristics of twenty-first-century conflict and the security threats facing Russia. Common themes in these articles were the likelihood of local and regional conflicts —rather than large-scale nuclear war between states, which was considered to be unlikely—as well as the threat to Russian national security from internal instability. Indeed, the possibilities of internal instability and regime change initiated by external actors were perceived to be a fundamental threat to national security. Another common theme was the lessons that could be drawn for Russia from the operations and interventions of the West (particularly the US and NATO) over the past decade. Writers such as Viktor A. Makhonin called for attention to be paid to the experience of all armed forces, not just the US, and stated that "modern wars are closely intertwined with nonmilitary forms and methods of confrontation."[8] In order to foresee changes in the character of armed conflict, some theorists argued that it was imperative to utilize the experiences of both Russian and other militaries.

This book investigates trends in Russian military thought and outlines the implications of Russian conclusions regarding the characteristics of contemporary and future conflict. It examines the debates around events that have shaped Russian thinking on the character of conflict and traces the evolution of this thinking in open-source material (particularly military journals), formal policy documents, and speeches. The issue of foresight and forecasting the future character of war is a common topic of discussion in the writings of serving or retired military officers in open-source publications such as *Voennaya mysl'*, *VPK, Vestnik Akademii voennyikh nauk*, and *Nezavisimoye voennoye obozrenie* (*NVO*), as well as in service journals such as *Armeiskii sbornik*. Analysis of these publications, which are intended for an internal Russian audience, provides new insights into what are considered to be key national security issues and facilitates a deeper understanding of how Russian military theorists perceive the character of conflict to be changing, the lessons they have learned from observing their military and others, the role of new technologies, new forms and methods, and the evolution of enduring methods such as the use of proxies. Prediction is an inherently difficult endeavor: attempting to extrapolate lessons and foresee future change on the basis of experience or the observation of other countries is problematic, constrained by bias, short termism, and path dependency.

Confirmation bias may contribute to a focus on examples that support existing views or beliefs, whereas examples that may challenge or contradict these views are ignored. Russia's failure to achieve a swift, decisive victory over Ukraine in February 2022 emphasized the significant gap between the theory and practice of war and the perils of failing to account for different contexts. Human behavior lies at the heart of war and conflict, increasing the unpredictability and uncertainty surrounding it. Attempts to impose a rigid scientific approach to understanding war may encourage misplaced confidence and the belief that all eventualities have been thought through. Nevertheless, it remains important to recognize the differing perspectives of others and the conclusions that they draw both from their own operational experience and that of others.

This book explores Russian military thought from both a historical and contemporary perspective to provide a greater understanding of the Russian approach to war. It looks at elements of both continuity and change in Russian military thought, tracing the legacies of Soviet-era concepts and writing in contemporary thinking. Observation of Western interventions and how Western states have used force in the post–Cold War era has been instrumental in shaping Russian thinking, as has Russia's own experiences. It examines the lessons learned from combat operations in Chechnya, Georgia, Ukraine, and Syria, as well as from Western military operations over the past two decades. Finally, it assesses whether this analysis has been translated into specific action in the post-Soviet era. Although the research and writing for this book was completed prior to Russia's invasion of Ukraine in February 2022, it would have been remiss not to include it. Thus, chapter 7 examines some of the early evidence from the war and the implications for Russian military thought. Key research questions that the book seeks to address include:

- What is the Russian view of the changing character of conflict and what future wars may look like?
- To what extent do Soviet concepts and military thought continue to influence contemporary thinking?
- How does Russia conceptualize military conflict?
- What lessons has Russia drawn from Western military activity since 1991?
- How well have lessons learnt during combat operations in Chechnya, Georgia, Ukraine, and Syria been incorporated into Russian military thought?
- How has technology influenced contemporary Russian military thought?

STRUCTURE OF THE BOOK

Chapter 1 explores the evolution of Soviet and Russian military thinking on the character of conflict, identifying continuities and change. It sets out the debates about how future wars might evolve alongside the enduring importance of military foresight and Russian focus on attempting to predict the character of future war as part of the field of military science. The need to prepare for future conflict reflects a theme in Russian strategic thought: anxiety, informed by history, about failing to predict a surprise attack or the action of an adversary. Russian military science, particularly its emphasis on foresight and forecasting the future character of war, is a deliberate countermeasure to this fear of surprise. This chapter also outlines how Russia conceptualizes military conflict and identifies enduring themes in the Russian view of the character of conflict.

Chapter 2 interrogates Russian analyses of Western interventions in the post–Cold War era and the lessons drawn from them, setting out concepts such as Major General Vladimir I. Slipchenko's "sixth-generation warfare," which was based on extensive analysis of the 1991 Gulf War and NATO's intervention in Serbia in 1999, and which appears to constitute a continuation and evolution of Ogarkov's thinking from the late Soviet era. Western interventions such as Operation Desert Storm in 1991, Operation Allied Force in 1999, and the ones in Iraq in 2003 and Libya in 2011 are frequently referenced in Russian theoretical military literature, used as important examples of the changing character of conflict in the twenty-first century.

Chapter 3 focuses on Russia's own military interventions in Chechnya, Georgia, Ukraine, and Syria, which are commonly discussed in the military theoretical literature. Since Vladimir Putin came to power in 2000, the Kremlin has demonstrated an increased willingness and ability to use the military lever to achieve broader strategic and foreign-policy goals. The 1990s were a period of turmoil and change for Russia. Putin took power when the country was perceived to be at its weakest, both domestically and internationally, encapsulated by the disastrous first attempt to quell separatism in Chechnya in 1994. Russia was initially unable to convert its extensive (numerically at least) military capabilities into military and strategic success, and thousands of Russian troops proved unable to secure the tiny republic. One of Putin's first priorities on taking power in 2000 was to halt the perceived decline of the Russian armed forces, which have undergone a comprehensive program of reform and modernization. The 2008 conflict with Georgia, the first Russian offensive operation against a foreign state since the end of the Cold War, demonstrated the renewed ability of the Russian armed forces to fight conventional wars following years of conflict in Chechnya and the North

Caucasus. Russian involvement in Syria has undoubtedly demonstrated that it is now able to project power beyond its own strategic backyard and that it is determined to play a global role.

Chapter 4 addresses Russian military thinking on the increasing role of information and communications technologies, drawing particular attention to the enduring Russian focus on the military aspects of conflict such as precision strike, again echoing Ogarkov's predictions from the 1980s. Western debates about the concept of network-centric warfare have directly influenced the direction of Russian military thought, leading to the development of an analogous Russian concept, network-centric warfare (*setetsentricheskaya voina*), which is considered to be both a key enabler and force multiplier. This chapter also examines the debate around the future role of emerging technologies, including artificial intelligence, robotics, and autonomous systems.

Chapter 5 examines Russian views on nonmilitary means of achieving strategic objectives and how Russia perceives other actors seek to achieve their strategic goals. In particular, it examines Russian concern about (and interest in) nonmilitary means of destabilization, such as "controlled chaos" (*upravliaemyi khaos*), including the perceived threat posed by color revolutions, which are conceptualized as distinct features of contemporary warfare. The views of soft power and "controlled chaos" as distinct features of contemporary and future wars are clearly expressed in the Russian military theoretical debate and are central to understanding the Russian view of contemporary conflict.

Chapter 6 considers one of the most prominent features of recent Russian military activity, the use of proxies and privatization of military forces. The use of proxy forces is not new: the USSR supported proxy forces worldwide during the Cold War, and Russia has used proxies to meet its goals across the post-Soviet space, from Transnistria to South Ossetia and Ukraine, since 1991. There has been a gradual evolution in its approach with the emergence of private military companies such as

Wagner in Syria. This chapter analyzes the roots of this evolution and explores the influence of Western activities.

Finally, chapter 7 examines initial evidence from the war in Ukraine and attempts to draw some preliminary findings for Russian military thought and the utility of military science's emphasis on foresight and prediction. It also examines the extent to which lessons learned from recent operational experience have been forgotten or even disregarded.

NOTES

1. A note on translations and abbreviations: transliterated Russian terms for words, phrases, and publications that are commonly associated with Russian military thought and the theoretical literature have been included.
2. Dima Adamsky, *The Culture of Military Innovation: The Impact of Cultural Factors on the Revolution in Military Affairs in Russia, the US, and Israel* (Stanford, CA: Stanford University Press, 2010).
3. Igor M. Popov and Musa M. Khamzatov, *Voina budushchego: kontseptualnye osnovy i prakticheskie vyvody* (Moscow: Kuchkogo Polye, 2017), 7.
4. W.R. Kintner and H. F. Scott, *The Nuclear Revolution in Soviet Military Affairs* (Norman: University of Oklahoma Press, 1968).
5. Sokolovsky used the term "revolution in military affairs" in his work on nuclear weapons and military strategy. Marshal Vasily D. Sokolovsky, ed., *Military Strategy: Soviet Doctrine and Concepts* (London: Pall Mall Press, 1963).
6. For a detailed analysis of the Soviet debates see Mary C. Fitzgerald, "Marshal Ogarkov and the new Soviet Revolution in Military Affairs," Research Memorandum CRM 87-2, January 1987, Center for Naval Analyses.
7. See Martin Van Creveld, *Technology and War: From 2000 BC to the Present* (London: Simon and Schuster, 2010); Steven Metz, *Strategy and the Revolution in Military Affairs: From Theory to Policy* (Collingdale: Diane Publishing, 1995); E. C. Sloan, *Revolution in Military Affairs* (Montreal: McGill-Queen's Press-MQUP, 2002).
8. Viktor A. Makhonin, "K voprosu o voennoi nauki i yee obekte," *Voennaya mysl'* no. 9, (September 2017): 26–36.

PART I

LESSONS LEARNED

CHAPTER 1

THE EVOLUTION OF
MILITARY THOUGHT

There is a long tradition of rigorous military strategic debate in Russia, as well as an emphasis on the systematic study of the theoretical foundations of war and conflict. Russian military thought is highly structured and takes a scientific approach, often combining empirical evidence with detailed mathematical calculations. The issue of foresight and forecasting the future character of war is an enduring concern, reflected in the writings of serving or retired military officers in open-source publications, many of which focus on the characteristics of twenty-first-century conflict and the security threats facing Russia, particularly those believed to derive from the US. A complex process of assessment, foresight, and forecasting has been in place since Soviet times, part of efforts to gain some illusion of control over the uncertainty and unpredictability of war and conflict. The enduring role of foresight and military science was emphasized in 2013 with the publication of a well-known article "The Value of Science in Foresight" by CGS General Valery V. Gerasimov. The piece represented an appeal to the Russian military science community to ensure the country was prepared for the challenges of conflict in the twenty-first century and posed some important questions: what constituted conflict and war in the contemporary era, had there been changes to the character of conflict and therefore the "forms and methods" (ways and means) used by armed forces, and what were the implications for Russia? Gerasimov emphasized the critical role that military science had to play, particularly in the development of a holistic understanding of asymmetric forms and methods, arguing that Russia "should not copy other people's experience and catch up with the leading countries, but work ahead of the curve and be in a leading position."[1]

This chapter explores Russian views of the changing character of conflict, the debates about how wars of the future may change, and the enduring focus of Russian military science on predicting the character of future war. It examines elements of both continuity and change in Russian military thought, tracing the legacies of Soviet-era concepts and writing in contemporary thinking. In the post-Soviet era, the debate among Russian military theorists about the definition of war and its character was renewed, and this chapter outlines how Russia conceptualizes military conflict, the trends and debates in Russian military thought, and enduring themes in the Russian view of the character of conflict, discussing the implications of Russian conclusions regarding the characteristics of contemporary and future conflict. It also identifies some of the key individuals who have written prolifically on the character of conflict, as well as the institutions that are tasked with examining it. The focus is on ideas, rather than capabilities, and the chapter seeks to encourage a greater understanding of the range of perspectives a peer competitor holds and the particular analytical processes that take place in an attempt to avoid mirror-imaging.

FORECASTING FUTURE WAR

The shape of future war, what it may look like, and the means that an adversary may use are a particular concern of Russian military theorists. Kharis I. Saifetdinov maintains that the "most important condition for foreseeing the character of armed conflict is the creation of a creative environment in the field of military scientific thought and tolerance of different views."[2] Foresight (*predvidenie*) and prediction (*predskazanie*) are complex analytical processes that combine empirical approaches, including assessment of previous conflicts, with more scientific ones, including the creation of models of future conflict and algorithms of potential enemy actions.[3] There is a critical distinction between foresight and prediction identified by Jacob W. Kipp, who noted that prediction "implies a determined outcome without requiring any action by the subject," whereas foresight is a "tool or weapon used by the subject to act upon the objective world."[4] The Russian military encyclopedia defines foresight as an "assumption, prediction or forecast" of changes in future military theory and practice, characterizing it as reliant upon knowledge of the laws of war and armed struggle and upon the use of general scientific methods and approaches specific to military science.[5] This definition also acknowledges that prediction plays a role in foresight.

The Russian focus on foresight and prediction vis-à-vis military strategic planning is a legacy of the Soviet era, when they were considered to be a fundamental part of the scientific approach to military thought and planning that attempted to look forward rather than make predictions based solely on events of the past, assessing long-term trends and possibilities in determining the risks to take in the short term.[6] Writing in the 1960s, Vasily Sokolovsky stressed the importance of military foresight and thinking about the character of future war, which, in his view, constituted a critical element of military strategy. He identified a number of determinants that would shape future wars, including the distribution of military and political forces; the quality and quantity of war material; and the military and economic potential, probable composition, and potential of opposing coalitions, as well as their geographical deployment.[7] Without a comprehensive assessment of what future wars and conflicts may look like, it is difficult for states to correctly prepare and organize their armed forces, equipment, and policy.

Thus, the debate about whether the character of conflict is changing is not just a philosophical pursuit.

Bolshevik military doctrine was balanced between a reliance on foresight and the imperative of paying heed to the unpredictability of chance (*sluchainost'*) and the dangers of provocation (*provokatsiia*).[8] Marxism-Leninism was deemed to be the basis for foresight, and the Soviet leadership was considered to have been bestowed with the ability to predict the future. Despite the damning evidence of 1941 and the Soviet leadership's failure to act upon intelligence of an impending German invasion, a number of Soviet generals credited Stalin with the gift of foresight: Major General Illarion I. Fomichenko claimed that the "ability to foresee is the strongest sign of Stalin's genius," and Major General Mikhail A. Isayev stated that "Stalin's strategic foresight and penetration, his ability promptly to divine and foil the enemy's designs...were brought out with exceptional force in the Great Patriotic War [World War II]."[9]

The German invasion in 1941 highlights one of the principal drivers for the focus on foresight and prediction vis-à-vis future war: an enduring concern about being taken by surprise, together with the desire to ensure that Russia has superiority and can maintain the initiative over an adversary during the initial phase of war. Understanding how an adversary may act or react, through observation and analysis of their use of force, enables an actor to either take defensive or offensive measures, and amplify the element of surprise. Concern about being taken by surprise also contributes to a desire to ensure superiority during the initial period of conflict and an emphasis on the use of deception (and surprise), making it difficult for an adversary to react in a timely fashion. Gerasimov has referenced the Soviet military theorist Georgii Isserson and his 1940 book *New Forms of Warfare*, in which the latter warned that "in general, war is not declared. It simply begins by earlier deployed forces. Mobilisation and concentration [of forces] refers not to the period after the declaration of a state of war...but [what took place] unnoticed, gradually, long before this."[10] Isserson was writing prior to the German invasion of 1941, but nevertheless his work still reflects concerns about being caught unaware by an adversary's preparation for offensive operations, an enduring concern of the Russian strategic community. This leads to a focus on surprise (*vnezapnost*), both in terms of not being taken by surprise, as has happened in the past, as well as the need to use surprise against adversaries in order to gain the initiative.

This is also associated with an enduring focus on time as a critical factor in armed conflict, resulting in significant weight being placed on decisive action during the initial phase of war, high operational tempo (facilitated by rapid mobilization and good command and control), and the seizing of the strategic initiative, often through preemptive action. The initial period of war (IPW) has long been a particular preoccupation of Russian military theorists, who have conducted a number of studies of the time from when hostilities are perceived to have commenced (even if covertly) until initial objectives are considered to have been achieved. Thus, the IPW is key to understanding the Russian view of war and armed conflict. The Soviet focus on seizing and maintaining the strategic initiative during the initial phase of hostilities was linked to Bolshevik thought and the need to prevent opponents from imposing their will on the country: if it did not seize the initiative and destroy its adversary, then its adversary will seize it and destroy it.[11] This reflects the logic of *kto kogo*—"who will prevail over whom"—a Soviet-era belief in the zero-sum nature of international relations. The 1940 Field Regulations underscore the Soviet emphasis on ensuring that it was not taken by surprise by an opponent, stating that "elements of the Red Army must never be caught unawares and must answer any surprise of the enemy with a decisive blow."[12] The advent of modern weapons—particularly nuclear—significantly increased the spatial scope of a war while simultaneously compressing time scales, making the initial period of war of "decisive importance to the outcome of the entire war."[13] Consequently it was seen as imperative to develop methods and weapons that enable the attainment of victory within the shortest possible timeframe and with minimum loss. In the twenty-first century, technological advances have continued to compress time scales, hence an enduring focus on the IPW. Vyacheslav V. Kruglov and Aleksei S. Shubin assert that in the twenty-first century the effective use of time has become a key determinant of victory in an armed conflict; Colonel-General Vladimir B. Zarudnitsky maintains that victory in future wars will be contingent upon quickly gaining superiority over an adversary and holding the initiative.[14]

The focus on trends in warfare that are developing and that affect how future war might unfold is part of the broader field of military science (*voennaya nauka*), which developed during the "golden age" of Soviet military thought in the 1920s and 1930s, driven by a failure to recognize incremental changes that had occurred in warfare since the Napoleonic era. Defeat by the Japanese in 1905 and failures during World War I prompted a recognition of the imperative to question outdated ideas about war and warfare.[15] The development of Soviet military thought during the 1920s and 1930s was influenced by both specific individuals, such as Mikhail Frunze, Mikhail Tukhachevskii, Vladimir Triandafillov, Georgii Isserson, and Aleksandr Svechin, and military experience, most notably the Russian Civil War. This pattern continues today: the observations and experiences of both foreign and Russian armed forces by individual military theorists shape broader debates regarding the use of force and the ways and means an actor utilizes in order to achieve their strategic objectives.

Understanding the nature of future war thus lies at the heart of military science, defined as the "system of knowledge concerned with the laws and military-strategic nature of war, the organizational development and preparation of the armed forces and the nation for war and the methods of waging it"—in other words, it is a "systemic science of war."[16] The fundamental objective of military science is to develop an understanding of war and seek to predict the character and course of future wars by providing a clear scientific approach to the preparation of future conflicts based upon sound military theory. The development of military science as a discipline in the USSR was rooted in the dialectical materialism that underpinned Marxism-Leninism, particularly the notion of the dialectic, which emphasizes both coherence and constant change, as well as interconnectedness and interdependence.[17] Change (progression) prompts a reaction, which engenders further change, and so on. Dialectical materialism dictates that a series of incremental changes could gradually accumulate to cause a sudden breakthrough: Soviet analysts were cautioned against extrapolating on the basis of trends and urged to identify points at which sufficient quantity would stimulate a qualitative shift. A final principle of dialectical materialism assumed the "negation of the negation," (i.e., development never advances along a straight line): "one trend (thesis) as it asserts itself is the dominant one, leading to the emergence of a counter-trend (antithesis) which negates the first, leading in turn to a final negation of the negation and a new trend (synthesis)."[18] According to Kipp, Ogarkov applied this principle to his 1982 analysis of trends in the development of military art, which led to the idea of a military-technical revolution. The Soviet approach to foresight was also shaped by a firm belief in the reciprocal relationship between theory and practice, both of which were seen to inform foresight.

Frunze was fundamental in the development of a specific Soviet military science, arguing in the 1920s for an agenda for military scientific research focused on "a clear and exact notion of the character of future wars;...a correct and exact calculation of those forces and means...at the disposal of our possible opponents; and...a similar calculation of our own resources."[19] According to James J. Schneider's study of Soviet military science, Frunze's analysis provided "the underlying methodology that determines the adequacy and soundness of any military theory. Without this...military theory is nothing more than military opinion. Military science, through the careful observation and scrutiny of military fact, builds a comprehensive and coherent causal description of the nature of future war."[20] Military history also plays an important role, feeding into military science and providing empirical evidence that supports the development of a concept of future war.[21]

Vladimir K. Triandafillov was one of the first Russian military theorists to seek to incorporate the concept of possible future war into the work of foresight; his book *The Nature of the Operations of Contemporary Armies*, published in 1929, has been described as a model for the method of engaging in foresight in military affairs.[22] He analyzed the impact of technological, sociopolitical, and economic change on militaries across Europe in the wake of World War I, in an attempt to understand what military operations in a future war might look like:

> One must be fully aware both of contemporary achievements in military equipment and trends in the further development of every type of weapon. Otherwise, one cannot understand those changes that may occur in the organisation of armies in the near future.[23]

Other military theorists of this era, notably Georgii Isserson and Aleksandr Svechin, also focused on the nature of future war and its implications for Soviet strategy. Writing in 1926, Svechin pointed out the impossibility of foreseeing the actual course of events during a war but emphasized the necessity of trying to understand the "phenomenon of war."[24] In 1940, Isserson explored the Spanish Civil War and Germany's 1939 invasion of Poland for insights into the emerging character of war, assessing how the forms and methods of waging war change. He maintained that all military literature after 1918 had been "devoted to the study and prognostication of the nature of a future war." His work was intended to study "new forms of struggle" in action and, like Triandafillov, he set out the impact of technological advances on military thought and operations, stating that "the forms and methods of waging war are always the product of political, economic, geographical, technical and other conditions in which the war arises and is conducted."[25]

The Soviets distinguished between military science,[26] the system of knowledge about the character and laws of war, and military art (*voennoye iskusstvo*), which covers the theory and practice of military operations (Russians today also make this distinction).[27] Military art is seen as lying at the core of military science; that is, it is a branch of military science and includes the application of military knowledge to achieve certain effects, necessitating independent and/or creative thought in officers. Thus, according to the Russian understanding of the term, military art is the "theory and practice of preparing for and conducting military operations."[28] It constitutes the intersection of theoretical knowledge and practical preparation for combat:

> Military art begins where, on the one hand, deep theoretical knowledge and their creative application help the commander to better see the general connection of the phenomena occurring and more confidently orient himself in the situation, and on the other hand, he, without being constrained by a general theoretical scheme, seeks to penetrate deeper into the essence of the actual situation, to grasp its advantageous and disadvantageous features and based on their analysis to find original solutions and methods of action that best suit the given specific conditions and the assigned combat mission.[29]

There is also an important distinction between "military thought" and "military science." Popov and Khamzatov argue that although the two terms are linked, they are not synonymous: military thought is defined as the "fruit of lengthy reflection" focused on the future rather than the analysis of past wars, differentiating it from military science. Military thought, they write, is comprised of "innovative, unexpected ideas that question the traditional basis of classical military art, resisting the stereotypes of military science." They compare the two to the difference between design ideas and engineering details. On its own, military thought is nothing more than interesting and logical deductions: new ideas and approaches to military theory must be understood, evaluated, interpreted, developed, substantiated, and put into practice, which is where military science is key. Thus, the issue of what future conflict will look like is a fundamental question for military science, a question that military thought can help solve.[30]

Despite agreement about the continued relevance of forecasting and military science in the post-Soviet period, there was still an active debate about the state of Russian military science and whether it was continuing to meet the country's needs. General Makhmut A. Gareev was critical of military science for failing to anticipate the far-reaching consequences of the economic burden imposed by the Cold War.[31] Speaking in 2004, Slipchenko also criticized contemporary military science in Russia, arguing that there was a failure to understand the wars of the future and address the key questions of what kind of wars await Russia in the future and what it should prepare for:

> If you open *Voennaya Mysl'*, which mirrors our military scientific thought, you will not find profound formulations of wars of the future...[T]hey mostly beat the dead horse of past generations of wars that took place in the world or in Russia. They rehash everything that is gone.[32]

A retired senior military officer who taught at the Academy of the General Staff, Slipchenko[33] focused on developing the discipline of forecasting the character of future wars. Gerasimov has also been critical of the state of Russian military science in the post–Cold War era, stating that it was no longer comparable with the "heyday" of military theoretical thought, the 1930s. He condemns what he perceived to be "disdain for new ideas, non-standard approaches, and a different point of view in military science," as well as an unacceptable "dismissive attitude to science on the part of practitioners."[34] Sergei G. Chekinov and Sergei A. Bogdanov have both written extensively on future war and forecasting. In a 2014 article they noted that although military forecasting had become a widely recognized subsection of military science by the end of the twentieth century, it was still not able to provide clear answers regarding the future shape of war and had failed to forecast apparent changes to the character of conflict displayed during the 1991 Gulf War:

> The greatest achievements in forecasting methods were obtained in determining quantitative indicators, carrying out opera-tional-strategic and tactical calculations, but the development of a methodology for long-term military forecasting associated with

the use of qualitative concepts and indicators is clearly lagging behind.[35]

They called for the development of military "futurology," defined as a "predictive science of the future," to address the contemporary challenges to Russia's national security and avoid previous mistakes. Key to this concept is the work done on military futurology by Ivan Vorobyev during the 1990s and 2000s, including a book he coauthored with Vyacheslav V. Kruglov and Aleksandr Suptel.[36] Vorobyev returned to the topic of military futurology and the failures of forecasting in 2008, arguing that military science had failed to predict the 1991 Gulf War and 2003 Iraq War because of the forms and methods used to wage them. He set out what he saw as the failures of forecasting, arguing that "previously developed methods of forecasting military events are far from perfect.... [I]t is necessary to focus on intuitive foresight and find more effective methods of forecasting."[37]

In 2016 Vyacheslav V. Kruglov asked why military science was still struggling to carry out the most fundamental of its tasks, forecasting future war. These theorists were not alone in voicing concern about the state of military science and forecasting.[38] Some argued that the national political and economic crisis experienced during the 1990s inevitably impacted military science, which no longer met "modern requirements."[39] According to this line of argument, Russian military science, once considered the most advanced in the world, had not been a priority for the Russian leadership following the disintegration of the USSR, leading to a theoretical and methodological vacuum vis-à-vis the study of armed conflict. Others suggested that changes in the character of contemporary conflict had challenged some of the fundamental concepts supporting Soviet military science, undermining its utility.[40] Ignat S. Danilenko set out what he saw as a global crisis in military science, which in turn reflected a crisis in the functional role of war in the post–Cold War era:

> Soviet military science adhered to a limited view of the essence of war, viewing it as a process of mass armed struggle, and therefore it was unable to master all the trends in the rapid evolution of war in the twentieth century; in particular, it ignored its unconventional direction. Consequently, it did not become a true science of war.[41]

According to this view, investigating the role of nonmilitary means is an important task within the realm of military science, despite its principal focus being armed conflict and the military means of waging it. This was accentuated by Mikhail P. Stepshin and Andrei N. Anikonov's 2021 analysis of the challenges of forecasting the character of future conflict and war: they proposed studying the development of new weapons and military equipment for clues as to the character of future conflict, emphasizing the key role of technology and technological advance.[42]

The Soviet-era processes of military science and forecasting are still widely adhered to,[43] with broad trends in military developments and warfare discussed in journals such as *Voennaya Mysl* (a Ministry of Defense publication) and *Vestnik Akademii Voennykh Nauk* (the Academy of Military Science's journal); this adherence feeds into predictions about the potential character of future war. Once the broad parameters of what future war may look like have been established, military strategy is then devised to bring together the necessary ways and means (forms and methods in Russian). Strategic documents such as Russia's Military Doctrine, which are revised and updated frequently, formally set out the state's view of the character of conflict and its threat perceptions, providing a useful snapshot of official thinking about the parameters of war and conflict at a given moment in time. As Gareev stated in 2017, it is important "to foresee in advance changes in the nature of armed struggle" in order to facilitate necessary organizational, doctrinal, and procurement changes.[44] He underscored the importance of learning from both the mistakes and successes that Russia had experienced, as well as from the experience of the US, China, and other armed forces. Chekinov and Bogdanov have also reiterated the continued relevance of forecasting in order to be able to understand how new weapons systems could be employed:

> Forecasting is a way to gain an insight into a situation in which employment of weapons based on new physical properties—new weapons having greater destructive power, longer range, higher accuracy and rate of fire, broader capabilities of reconnaissance and robot-controlled assets, automated weapons control, communication, and information warfare, and closer integration of space-based, aerial, and ground reconnaissance systems in target designation and acquisition in real time—will have a significant impact on the fast pace of future wars.[45]

The principal institutions involved in analyzing international military trends, predicting the likely parameters of future wars and their implications for Russia are the General Staff, including its think-tank the Center for Military-Strategic Studies (*Tsentr Voenno-Strategicheskikh Issledovanii Generalnogo Shtaba Vooruzhennykh Sil' Rossiyskoi Federatsii*), and the Academy of Military Science (AVN). The Center for Military-Strategic Studies was set up in 1985 to conduct applied research on issues related to strategy, operational art, and the effects of changes in the character of conflict, including the impact of new technologies. It attempts to foresee potential future threats to Russian national security over a period of thirty to fifty years and predict the shape of future war. According to its director, Major General Aleksandr Smolvy, the center remains an important source of ideas in military science, researching the character of military conflicts to develop a "system of forms and methods of action of both a military and nonmilitary nature." A particular focus of the center is analyzing the experience of both foreign actors and Russia in contemporary military conflict in order to draw lessons for the future deployment of the Russian armed forces. Smolvy describes the Center for Military-Strategic Studies as a unique organization, which fuses military science with practical experience through its interaction with other Ministry of Defense (MoD) organizations, research institutions, the armed forces, and the defense industry.[46] Some of the most prominent voices in Russian military science and forecasting have spent time at the center, including Sergei G. Chekinov and Sergei A. Bogdanov, both of whom have headed it up.

The AVN was established in 1994 as a forum for military scientists and other security experts to analyze contemporary conflict, debate the theory of war, engage in forecasting and prediction, and make practical recommendations for the Russian armed forces.[47] General Makhmut Gareev, the first president of the AVN, believed that foresight and predicting newly developing methods of war had endured as one of the most important tasks of military science. Describing foresight as similar to the labors of Sisyphus, he stated that it had become ever more complicated as the "qualitative disparity increases between weapons of the last war and those of the future wars" and warned that the "next war tends not to resemble the previous one but throws up progressively more novelties."[48] Gareev, who died in 2019, was succeeded as president of the AVN in 2021 by General Gerasimov. As a result, Gerasimov will continue to play a central role in the development of Russian military thought and the discipline of military science. Since becoming the Russian Chief of the General Staff in 2012, he has used his annual address to the AVN to set out his views on the future of warfare and appeal to Russian military scientists and other experts to focus on critical issues. Speaking in 2017, he concluded that, although the substance of wars in the present and foreseeable future will remain the same, characterized by armed struggle, the definition of war and conflict will continue to evolve, necessitating "careful study":

> An urgent task is the formation of scenarios, long-term forecasts of the development of the military-political and strategic situation in the most important regions of the world. It is necessary to quickly study the features of modern armed conflicts...[and] develop methods for the operation of military command and control and actions of troops in various conditions.[49]

In his view, such study is critical to national defense, and he concluded by emphasizing the enduring relevance of military science. During his time as CGS, Gerasimov has sought to reinvigorate Russian military science, issuing numerous "calls to arms" for military scientists and theorists. Writing in the journal of the AVN in 2019, he stated that new arenas of confrontation in contemporary conflicts had resulted in the increasing use of nonmilitary means—including political, economic, and information tools, supported by military force—and declared that those involved in formulating military strategy[50] should be seeking to predict the character of future conflict in order to ensure that the Russian armed forces were being trained and equipped to meet the challenges of the future. He accentuated the importance of adaptation and called for military scientists to increase their research on new methods of using advanced weapons, as well as how to counteract the possibility of military actions by an adversary both in space and from space.[51] In his view, digital technologies, artificial intelligence, unmanned systems, and electronic warfare needed to feature prominently on the agenda for the future development of military science. Thus, the focus was on advanced technologies, rather than intangibles such as leadership and the quality and effectiveness of training and morale.

Tor Bukkvoll has identified three camps of Russian military theorists: traditionalists, modernists, and revolutionaries.[52] In his view, traditionalists (typified by General Makhmut A. Gareev) tend to have more conservative approaches to warfare and emphasize the enduring relevance of historical experience, particularly the Great Patriotic War. Gareev cautioned about constantly looking for a revolutionary approach to warfare, citing Polish philosopher Leszek Kolakowski:

> We need to remember two things: first, if not for new generations' rebellions against old traditions, we would have been still living in caves; and second, if there had only been those rebellions, we would have been in the caves again.[53]

Table 1. Key contributors to Russian military science debates.

Name	Role	Service Experience
General Valery Gerasimov	Chief of the General Staff; head of the AVN since 2021	Joined Soviet Armed Forces in 1970s (infantry, tank command) Commanded 58th Army and 144th Guards Motor Rifle Division. Appointed Chief of the General Staff 2012.
General Makhmut Gareev	President of the AVN, 1995–2019	Joined Soviet Armed Forces in 1941 (infantry), fought in Great Patriotic War. Deputy Soviet Chief of the General Staff (Nikolai Ogarkov was CGS at the time) Retired 1992.
Major General Vladimir Slipchenko	Member of the AVN	Served in Soviet Armed Forces for over 40 years (artillery), left in 1993.
Colonel Sergei Chekinov	Worked at TsVI GSh; director from 2009	Served in the Air Defence Forces and Air Force.
Lieutenant General Sergei Bogdanov	Head of TsVI GSh, 1990–1995. Member of the AVN	Served in Soviet Armed Forces since 1960s (armoured forces; graduated Ulyanovsk Tank School 1963); General Staff since 1980.
Major General Ivan Vorobyev	Member of the AVN	Joined Soviet Armed Forces in 1940, fought in Great Patriotic War; infantry. Taught at Frunze Military Academy; professor at Combined Arms Academy.
Colonel Valery Kiselyev	Member of the AVN	Joined Soviet Armed Forces in 1974, commanded motor rifle regiment. In 2000 became professor at Frunze Military Academy; from 2003, at Combined Arms Academy.
Colonel General Vladimir Zarudnitsky	Head of the Military Academy of the General Staff of the Russian Armed Forces	Joined Soviet Armed Forces in 1975. Head of the Main Operational Directorate 2011–2014, commander of the Central Military District 2014–2017.

Modernists adopt an approach that facilitates the modernization of doctrine, tactics, weapons, and equipment, while recognizing the continuing importance of historical experience. Finally, revolutionaries emphasize the need for new approaches and a break with the past. This camp is epitomized by Slipchenko, whose principal ideas are discussed in detail in the next chapter.

THE RUSSIAN UNDERSTANDING OF WAR AND CONFLICT

There is an ongoing debate among Russian theorists and policy makers about what actually constitutes war and conflict; a number of experts perceived the character of conflict to be changing, driven largely by an increase in the utilization of nonmilitary means and methods. The rapid evolution of armed conflict in an apparently unconventional direction means that military scientists have focused on investigating the role of nonmilitary means within conflict, despite the principal focus on military science being armed conflict and military means of waging war. Russian experts believe that there has been a change in the forms and methods that states use to wage war in both the military and nonmilitary realms. Speaking at the Valdai discussion club in October 2021, Vladimir Putin outlined his views of how conflict in the twenty-first century has changed: "Previously, a war lost by one side meant victory for the other side, which took responsibility for what was happening...Things are different now: no matter who takes the upper hand, the war does not stop, but just changes form."[54] This highlights a prevalent view that war is a continuum of confrontation, which may or may not take a violent form, reflecting the Soviet view of war and conflict: "if war is a continuation of politics, only by other means, so also peace is a continuation of struggle only by other means."[55]

Garthoff notes that the Soviet view of war and conflict went far beyond the Clausewitzian understanding, presuming "permanent conflict (although not necessarily armed), even in peace."[56] According to this understanding of perpetual conflict, the distinction between war and peace is eradicated, leaving a continuous spectrum with peace and war at opposing ends; the only difference between the two is the extent to which armed force is employed. The Soviet understanding of war was fundamentally shaped by the principles of Marxism-Leninism, which characterized war as a continuation of politics by means of armed violence, equating it with both politics and the class struggle as a whole.[57] Lenin's conviction was that any war is intrinsically linked to the political system that it emanates from: war was viewed as the apogee of the ideological struggle between the USSR and its class enemies, a conflict in which the pursuit and achievement of victory, rather than denying an opponent victory or terminating conflict, was the principal objective.[58] Compromise was not considered to be an option, either in external relations or internally—Garthoff described this as a "combat frame of reference" that was applied to political relations, summed up by the notion of "destroy or be destroyed," reflecting the zero-sum notion of *kto kogo.*[59] Thus, Soviet military thought focused on a narrow ideological perspective: fighting an uncompromising, decisive war in order to defeat its capitalist adversary and ensure that socialism prevailed worldwide.[60] There was a brief period during the Great Patriotic War when this ideological approach to war was perceived to have disappeared. The German invasion in 1941 undermined the argument that the USSR's principal threat arose from the class struggle because the enemy had nothing to do with class but posed an existential threat to the state.[61]

Svechin was a consistent advocate of the centrality of politics within military operations (echoing Clausewitz's understanding of the purpose of war); at the same time, however, he advocated the idea of the reasonable sovereignty of each in the sphere of their competency. Writing in the 1920s, he was highly critical of those military leaders and theorists who advocated the independence of military strategy from politics: in his view, strategy cannot exist in a vacuum without politics and is condemned to pay for all the sins of politics:

> War is only a part of political conflict....[B]ecause the leaders
> of military operations are responsible for only a part, albeit an
> essential part of this political solution, they must be subordinate
> to political requirements.[62]

The Soviets had a firm belief in the utility of the military lever of power in support of political objectives, but there was also a focus on nonmilitary means, driven in part by observation of other states. Sokolovsky noted in the 1960s that a number of foreign military publications were focused not just on the military means of conflict but also the nonmilitary means, such as ideological, political, psychological, economic, and diplomatic levers of influence. He referenced Basil Liddell Hart's assertion that the means of war include not only the armed forces, but nonmilitary means, including economic pressure, propaganda, diplomacy, and subversion.[63] The contemporary focus of Russian military theorists and thinkers such as Gerasimov and Gareev on the role of nonmilitary means in conflict, and the blurring of the line between war and peace, is therefore nothing new; it has long been a preoccupation. The growing use of nonmilitary means to achieve strategic objectives has been the subject of much discussion, leading to questions about the character of conflict and what constitutes war in the twenty-first century. A number of Russian experts perceive the character of conflict to be changing, driven largely by an increase in the utilization of nonmilitary means and methods: there is a widespread belief that contemporary wars are dominated by strategies of indirect, asymmetric action that use a combination of military and nonmilitary means to influence an adversary with the intention of undermining their will to resist.

This has prompted a debate about whether nonmilitary means of waging war can be defined as "war" (*voina*) or needed to be classified as a "struggle" (*bor'ba*). A particular focus of debate has centered around whether war must always involve armed violence. Military actions still constitute the essence of war, according to a number of Russian experts and theorists, such as Aleksandr I. Vladimirov, who is highly critical of those who believe that the nature of contemporary war has changed, arguing that violence is the very essence of war; in his opinion, what has changed is the character of war, its aims, how it is conducted, and its forms and methods.[64] In 2017 Russian CGS Valery V. Gerasimov drew attention to the long-running debate between experts on the definition of war, noting that some adhered to a Clausewitzian, classical interpretation of war as an "act of violence intended to compel an opponent to fulfil our will." Others were calling for a fundamental review of the term, because, in their view, armed violence was no longer a defining attribute of war: nonmilitary aspects such as economic and information means had become central to contemporary conflict.[65] The definition of war (*voina*) in the *Military Encyclopedic Dictionary* states that armed violence is the principal means of war, but it also references other forms of violence:

> The use of means of armed and other types of violence to achieve socio-political, economic, ideological, territorial, national, ethnic, religious, and other goals. The main content of war is armed struggle.[66]

In 2003, Gareev argued that armed violence constituted the main essence of war, stating that "war in its true sense is associated with military action. War using only nonmilitary means cannot be."[67] He reiterated this line of thought in 2015, noting that the growth in nonmilitary means of pursuing strategic objectives, such as economic, informational, cyber, and psychological means, has led to questions about the concept of war and whether it has changed:

> Confrontation without the use of weapon is a fight (*bor'ba*), whereas continuation of politics through violence, with the use of armed forces, is a war (*voina*). Some...have claimed that all non-military means are a recent phenomenon and they treat their use as a war. If the use of all non–military means in international rivalry is a war, then the whole history of mankind is a war indeed....War is a continuation of politics through violent means

(military actions) along with other forms and non–military means, so–called "soft power."[68]

In Gareev's view, war is just one of the means of pursuing a policy which, in the opinion of a number of Russian theorists, is an endless, uncompromising process of rivalry, confrontation, and fighting between different actors in pursuit of their own ends, which may sometimes be supplemented with military action. Makhonin argues that modern wars are "closely intertwined with nonmilitary forms and methods of conflict," and Vladimir V. Babich takes a similar view, asserting that confrontation is a natural part of human existence: during peacetime, this confrontation takes place in the nonmilitary arena, such as the economic or information realms, but during war, nonmilitary means are substituted by military action.[69] This reflects a bleak view of politics and international relations as a permanent struggle for power and resources; war and peace are just different stages of the same process, different ends of the same continuum.

Gareev's characterization of war is not accepted by all and a number of Russian experts maintain that armed violence is not the defining feature of war, reasoning that in the twenty-first-century actors can achieve their strategic objectives by means of alternative forms of violence, such as with economic, informational, and psychological tools, obviating the need for direct armed intervention. Mikhail M. Kurochko rejects the notion that armed violence should be solely associated with the use of physical weapons, arguing that violence can be inflicted by other means and the nature of the weapon varies depending on the target. He defined the idea of "non-classical" war, which targets the consciousness of individuals and society using non-kinetic violence in an indirect way, such as informational-psychological or economic. He contrasts this with classical war in which the principal aim is the destruction and capture of physical assets, both individuals and materials.[70] Aleksandr Bartosh has written a number of articles on contemporary conflict, concluding that military violence—defined as the use of weapons to physically suppress an adversary and bend them to your will—constitutes the essence and defining feature of war.[71] Nonviolent means can be used to weaken an adversary and their military capabilities, but, in his view, "the final word remains on the battlefield." Viktor N. Gorbunov and Sergei A. Bogdanov argue that war is distinguished "not by the form of violence, but by its main features: an uncompromising struggle using the means of violence for a certain period of time, the victory of one of the parties and the defeat of the other."[72] A 2010 review of Russian military science and the character of war found that the majority of scholars favored using the term "war" when states deploy military force and wage an armed struggle. All other forms of struggle or confrontation, conducted during both peacetime and war, are defined by the term "struggle."[73] There is also agreement on the political nature of war: Mikhail Borchev states that war may not always be associated with armed struggle, but it is "always an exclusive means of achieving political plans."[74] This is echoed by Dmitry Baluyev, who believes that war is shifting from the military to the political sphere and who calls for better understanding of changes in the character of war, in particular its technological, information-psychological, and political aspects.[75] Popov and Khamzatov have conducted a wide-ranging analysis of the terms "war" and "conflict," using Clausewitz as a starting point. They criticize those who interpret Clausewitz's definition of war as simplistic and proceed to use an overly simplistic definition, arguing that the Prussian thinker never offered a single, unambiguous definition. Nevertheless, they return to Clausewitz in setting out a definition for war as "an act of violence intended to compel an opponent to fulfil our will," a definition echoed by Gerasimov in 2017, discussed earlier.[76] Gerasimov has also noted that there is no single definition of war in Russia's official policy documents: Russia's Military Doctrine defines war as a "form of solution for interstate or internal state disputes through the use of military force."[77]

The debate about what constitutes "war" and whether nonmilitary means fall within accepted definitions of the term reflect perceived changes to the character of conflict in the twenty-first century. The general view of both military theorists and policy makers in the early twenty-first century was that the character of conflict was undergoing change. From 2008 there was a consistent number of articles in Russian military publications such as *Voennoyaya Mysl'* and *Voenno-promyishlenniyi kur'er* focusing on the characteristics of twenty-first-century conflict and the threats facing Russia, particularly from the US.[78] There was a belief that Russia needed to be prepared for multiple future conflict scenarios across the spectrum of activity, with two opposing themes that can be identified in the literature and policy documents: the first is the role and impact of new technologies (such as artificial intelligence, nanotechnology, weapons of new physical principles) and high-precision weapons. V. Bocharnikov, S. V. Lemshev, and G. V. Lyutkene identify a qualitative change in the means of warfare, driven by the development of precision weapons, alongside a compression of time, both in terms of "preparing, unleashing and waging wars."[79] Secondly, there is a focus on, and concern about, the increasing use of nonmilitary means by state actors to achieve their strategic objectives, including the perceived threat to national security from externally sponsored protests, subversion, and regime change (particularly sponsored by the US), including color revolutions.[80] Vladimir I. Lutovinov divides nonmilitary means into four distinct categories: political and diplomatic means; legislative (complex activities aiming at manipulating norms of international law); economic (destabilizing an adversary's economic potential); and information-psychological and spiritual means. In his view, the latter category is the most important because it can be used to weaken a country from within.[81] Gorbunov and Bogdanov outlined the appearance of new forms of conflict in which military means are either not used at all or have no fixed role, arguing that the weakening of a state in order to deprive it of the will to resist is a key objective of contemporary aggression.[82]

Bogdanov developed this idea further in a 2013 article with Chekinov on what they termed "new-generation war," which emphasized the high importance of asymmetric actions aimed at neutralizing the enemy's military superiority through the combined use of political, economic, technological, ecological, and information campaigns. In their view, "decisive battles will rage in the information environment"; again, the focus is on nonmilitary means, reflecting the Russian view that in contemporary warfare the principal battlespace is the mind. Chekinov and Bogdanov outlined two phases of war: the opening phase, which is the longest, and the closing phase. In their view, the opening period starts with an extremely intensive, months-long coordinated nonmilitary campaign launched against the target country, including diplomatic, economic, ideological, psychological, and information measures. The closing phase, when military forces are deployed, should be much shorter because they are only deployed once a state's will to resist has been undermined during the opening phase. Chekinov and Bogdanov consider the first, predominantly nonmilitary phase of the conflict to be much more important than the second. The main objective is to reduce the need to deploy hard military power to the minimum necessary by undermining the adversary's will to resist.[83] Vladimir M. Moiseev argues that contemporary conflicts are characterized by the use of "more selective and less destructive means," intended not to destroy an adversary but to compel him to a particular course of action.[84] This is echoed by Vyacheslav V. Kruglov and Aleksei S. Shubin, who believe that there have been revolutionary changes in the means of warfare, driven by precision-guided munitions, which have compressed the critical factor of time and accentuated the importance of intelligence-driven preemptive action.[85]

Gerasimov has written widely on the changing character of conflict. In his view, contemporary conflicts differ from each other in terms of participants, weapons, and the forms and methods used. He also defines various types of struggle, including armed, as well as "political, diplomatic, informational and others." He has identified a number of features that are, in his opinion, characteristic of conflict in the twenty-first century, including the central role of information superiority, alongside the increasing use of artificial intelligence (AI), unmanned aerial vehicles (UAVs), and "diverse" forces, such as private military companies (PMSCs).

In addition to the impact of modern technologies on the character of contemporary conflict, he also identified several key characteristics of modern war: destabilization of a state through subversive action and information confrontation:

> Future wars will see a shift in hostilities into [the informa-tion] sphere...[I]nformation technology is becoming one of the most promising weapons. The information sphere, having no pronounced national borders, facilitates remote, covert influence not only on critical information infrastructures, but also on the population of the country, directly affecting the state of national security of the state. Consequently, the study of... information activities is the most important task of military science.[86]

Reflecting on Russia's experience in Syria, Gerasimov asserted that conflict in the twenty-first century involves the integrated use of nonmil-itary means, such as political, economic, and information tools, with the support of the armed forces. The "essence" of contemporary war is the pursuit of political goals through the application of minimal military pressure on an adversary, at the same time as destroying their military and economic potential, undermining their will to resist through the use of information-psychological influence and subversion: "information has become one of the most effective weapons."[87] In 2018, Gerasimov again set out his vision of future war, based on Russia's operational experience in Syria, as well as lessons learned from observing Western interventions:

> Every military conflict has its own distinctive features. The main features of future conflicts will be the widespread use of high-precision and other types of new weapons, including robotic ones. The economy and the system of state administration of the enemy will be priority targets. In addition to the traditional spheres of armed struggle, the information sphere and space will be actively involved.[88]

Gerasimov's views were endorsed by Aleksandr V. Dvornikov, the former commander of Russian forces in Syria, who asserted that the conflicts of the post–Cold War era (he directly referenced Yugoslavia, Iraq, Libya, Syria, and Ukraine) prove that the character of conflict is undergoing significant change. Whereas the conflicts of previous centuries centered around massed armies, in the twenty-first century the emphasis is on achieving strategic objectives by means of integrated forces, long-range precision strike, peacekeeping, and crisis management: "an 'obedient' government is established in the state, the country is fragmented, chaos and lawlessness are sown, control established over resources, and military bases of the aggressor established."[89]

Thus, by 2021, Russian military theorists had concluded that, although the role of nonmilitary means in conflict had increased, in particular psychological struggle,[90] the impact of technological advances on the forms and methods of war meant that there was still a substantial focus on military means. Stepshin and Anikonov argued that scientific and technical progress means that future armed conflict will acquire new characteristics associated with the use of military and nonmilitary measures, in addition to new weapons and military equipment.[91] The 2020 Nagorno-Karabakh war highlighted the significance of high-tech weapons and equipment, particularly the role of UAVs and networked systems for reconnaissance, command, and control, as well as the further compression of time during hostilities.[92] Attention has also turned to the concept of multi-domain operations, following analysis of Western documents and concepts such as anti-access area denial.[93] Ya A. Chizhevsky suggests that twenty-first century war is not limited in terms of time, space, or the number of protagonists and is conducted in all domains, at all levels (from the strategic to the tactical), and by all available means. He points to network-centric warfare and horizontal networks as being emblematic of contemporary armed conflict, facilitating superiority over an adversary because they offer far greater flexibility and faster decision-making than vertically integrated hierarchical structures, which are consequently more vulnerable.[94]

The evolution of theoretical debates and official concerns about the characteristics of contemporary conflict is evident in formal policy documents, particularly Russia's Military Doctrine. This is revised and updated frequently, formally setting out the state's perception of threat, its view of war, and the anticipated character of armed conflict in order to facilitate formal direction for the use of the armed forces. Thus, these documents all together show official thinking about war and conflict at certain time and how future wars might evolve. It is important to understand that there is a difference between Russian and Western usage of the term "doctrine." Western understanding tends to be focused on the principles that guide how military forces are deployed;[95] in contrast, the Russian definition is broader and encompasses views on the fundamental nature of war and its character. Sokolovsky defined military doctrine as an "expression of the accepted views of a state regarding the political evaluation of future war, the state attitude toward war, the definition of the nature of future war, preparation of the country for war..., the problems of forming and training the armed forces, as well as the methods of warfare."[96]

All iterations of the Russian Military Doctrine (and National Security Strategy) since 1993 have identified attempts to overthrow the constitutional system by force as one of the principal internal sources of military threat, although the language used has intensified with each iteration. The 1993 Military Doctrine identified "attempts to interfere in the internal affairs of and destabilize the internal political situation in the Russian Federation" as a source of external military danger, while noting that a significant source of internal military threat stemmed from the possibility of "attempts to overthrow the constitutional system by force or to disrupt the functioning of organs of state power and administration." The 2000 Military Doctrine reiterates the threat from attempts to overthrow the constitutional order by force but goes further in a section on the nature of modern wars, key features of which are thought to include the extensive use of "indirect, non-close-quarter, and other (including non-traditional) forms and means of operation," including long-range electronic engagement.

The 2010 iteration of the doctrine codified Russian views of the changing character of conflict and greater global instability, noting an intensification of military dangers to Russia, including "attempts to destabilize the situation in individual states and regions," as well as interference in a state's internal affairs. It also referred to the integrated utilization of military force with resources of a nonmilitary nature, along with the much greater role of information confrontation. Features of modern military conflicts were deemed to include unpredictability, a wide range of objectives, and the prior implementation of information warfare measures in pursuit of political objectives without the deployment of military force. The 2014 Russian Military Doctrine notes the growth in global competition and tensions, including "rivalry of proclaimed values and models of development" and the "tendency towards shifting military risks and threats to the information space and internal sphere of the Russian Federation," emphasizing the threat to vital national interests from foreign subversive influence.[97] Contemporary military conflict is characterized by the "integrated employment of military force and political, economic, informational and other nonmilitary measures implemented with widespread use of protest potential of population and Special Forces."[98] This echoes the language used by Gerasimov in his famous 2013 article published in *VPK*[99]: the protest potential of the population is characterized as a new tool of warfare, which can and will be exploited by adversaries, largely by subversive action.

Russian military thought is highly structured and takes a very scientific approach. As discussed earlier, military thought is an important component of Russian military science, providing the broad design ideas rather than the specific engineering details. There are long-running debates within the wider Russian military strategic community about the features of contemporary conflict. These form part of the enduring practice of foresight and forecasting, which seeks to understand what

constitutes conflict and war in the contemporary era, perceived changes
to the character of conflict and therefore the forms of methods used by
armed forces, and the implications for Russia. Military theorists draw
their conclusions from their own observation of the experience of foreign
(principally Western) armed forces, as well as the experience of Russia.
These conclusions may then be integrated with lessons from Russia's own
operational experience, leading to adaptation and, sometimes, emulation.

   The Russian understanding of war is complex; it is viewed as a sociopolitical, military phenomenon characterized by the diversity of means
available, as well as its multidimensionality. There are two opposing
themes that can be identified in the literature and policy documents: the
first is the role and impact of new technologies (AI, nanotechnology,
weapons of new physical principles) and precision-guided weapons.
The second is a focus on the increasing use of nonmilitary means by
state actors to achieve their strategic objectives, which prompts concern
about the perceived threat to Russian national security from internal
instability, subversion, and regime change initiated by external actors.
Although there are a number of continuities in Russian assumptions
about adversaries and how they seek to achieve their national objectives
using all available means, there have also been noticeable changes, which
this book is going to explore in more detail in subsequent chapters.
These changes include a significant focus on command and control
(particularly network-centric warfare), the establishment of the National
Defense Management Center (*Natsionalyi Tsentr Upravleniya Oborony*, or
NTsUO), the importance of information as a key element of the character
of conflict, and precision firepower.

NOTES

1. Valery V. Gerasimov, "Tsennost' nauki v predvidenii," *Voenno-promyshlennii kur'er,* February 2, 2013, https://vpk-news.ru/articles/1 4632?.
2. Kharis I. Saifetdinov, "Aleksandr Svechin: vydayushchiysya voyennyy myslitel' XX veka," *Voennaya mysl'* no. 8 (2018).
3. V. D. Ryabchuk and V. I. Nichipor, "O roli i meste prognozirovaniya i predvedeniya v sisteme planirovaniya operatsii I obshchevoiskogo boya," *Voennaya mysl'* no. 10 (October 2007): 61–67; See also V. D. Ryabchuk, "Problemyi voennoi nauki i voennogo prognozirovaniya v usloviyakh intellektual'no-informatsionnogo protivoborstva," *Voennaya mysl'* no. 5 (May 2008): 67–76.
4. Jacob W. Kipp, *The Methodology of Foresight and Forecasting in Soviet Military Affairs* (Fort Leavenworth, KS: Soviet Army Studies Office, 1988), 6.
5. Ministry of Defense of the Russian Federation, *Voennyyi entsiklopedicheskii slovar'*, https://encyclopedia.mil.ru/encyclopedia/dictionary.htm. For the dictionary definition of predvidenie (foresight), see https://encyclopedia.mil.ru/encyclopedia/dictionary/details.htm?id=9307@morfDictionary.
6. S. Kozlov, "K voprosu o razvitii sovetskoi voennoi nauki posle vtoroi mirovoi voiny," *Voennaya mysl'*, no. 2 (1964); Raymond L. Garthoff, *How Russia Makes War: Soviet Military Doctrine* (London: George Allen & Unwin, 1954), 11. The Soviet armed forces were generally deployed offensively in those situations where other, less risky methods were not considered feasible, but there was still substantial potential for advance. The armed forces remained the primary instrument for advancing Soviet aims, but they relied heavily on other tools, such as subversion, sabotage, insurgency, and proxy war.
7. Sokolovsky, ed., *Military Strategy,* 12.
8. Garthoff, *How Russia Makes War,* 256.
9. Cited in Garthoff, *How Russia Makes War,* 254.
10. For more on Isserson, see Steven J. Main, "'You Cannot Generate Ideas by Orders': The Continuing Importance of Studying Soviet Military History—G. S. Isserson and Russia's Current Geo-Political Stance," *The Journal of Slavic Military Studies* 29, no.1 (2016): 48–72.

11. Garthoff, *How Russia Makes War*, 88. For a wide-ranging analysis of the IPW, see Sergei A. Bogdanov, "Nachal'nyi period voinyi: istoriya i sovremennost," *Voennaya Mysl'* no. 11 (November 2004): 15–24.
12. Cited in Garthoff, *How Russia Makes War*, 275.
13. Sokolovsky, ed., *Military Strategy*, 204. For more on Russian assessments of the initial period of war see Stephen J. Cimbala, "The initial period of war: Russia's Soviet heritage," *The Journal of Slavic Military Studies* 15, no. 2 (2002): 59–88.
14. Vyacheslav V. Kruglov and Aleksei S. Shubin, "O vozrastayushchem znachenii uprezhdeniya protivnika v deistviyakh," *Voennaya mysl'* 12 (December 2021): 27–34; Vladimir B. Zarudnitsky, "Kharakter i soderzhaniye voennyikh konfliktov v sovremennykh usloviyakh i obozrimoi perspective," *Voennaya Mysl* 1 (January 2021): 34–44.
15. James Schneider, "Introduction," in Giorgii Isserson, *The Evolution of Operational Art*. Translated by Bruce W. Menning (Fort Leavenworth, KS: Combat Studies Institute Press, 2013), vii–viii.
16. Definition from the *Military Encyclopaedic Dictionary* quoted in Aleksandr I. Vladimirov, *Osnovyi obshchei teorii voinyi v 3 chastyakh. Chast 1: osnovyi teorii voinyi* (Moscow: Universitet Sinergiya, 2018), 249. It is also quoted in Ignat S. Danilenko, "Ot prikladnoi voennoi nauki – k sistemoi nauke o voine," *Voennaya Mysl* no. 10 (October 2008): 26. For a detailed analysis of the term "military science" see Makhonin, "K voprosu o voennoi nauki i yee obekte," 26–36.
17. Kipp, *The Methodology of Foresight and Forecasting in Soviet Military Affairs*, 4.
18. Kipp, 9.
19. Quoted in James J. Schneider, "The origins of Soviet military science," *The Journal of Soviet Military Studies* 2, no. 4 (1989): 514.
20. Schneider, 516.
21. Schneider, 516.
22. Kipp, *The Methodology of Foresight and Forecasting in Soviet Military Affairs*, 11.
23. Vladimir K. Triandafillov, *The Nature of the Operations of Modern Armies* (Abingdon: Routledge, 1994), 9.
24. Aleksandr Svechin, *Strategy*, translated from the Russian (Minneapolis: East View Publications, 1992).
25. Giorgii S. Isserson, *GS Isserson and the War of the Future*. Translated and edited by Richard W Harrison (Jefferson, NC: McFarland & Company, 2016).

26. D. M. Glantz has emphasized the difference between Soviet and Western understandings of the concept of "military science," maintaining that the US has neither a well-developed and focused body of military knowledge nor an analytical process that compared with Soviet military science. The US does not systematically study and critique its past military experiences and the past military experiences of other nations. D. M. Glantz, *Soviet Military Operational Art: In Pursuit of Deep Battle* (New York: Frank Cass, 1991), 2.
27. Derek Leebaert, "The Context of Soviet Military Thinking," in *Soviet Military Thinking*, ed. Derek Leebaert, *Soviet Military Thinking* (London: George Allen & Unwin, 1981), 14.
28. Ministry of Defense of the Russian Federation, *Voennyyi entsiklopedicheskii slovar'*, https://encyclopedia.mil.ru/encyclopedia/dictionary.htm.
29. Makhmut A. Gareev, "O vyirabotke u ofitserov kachestv i navyikov, neobkhodimyikh dlya proyavleniya vyisokogo urovnya voennogo iskusstva," *Voennaya mysl'* no. 12 (2017): 72–73.
30. Popov and Khamzatov, *Voina budushchego: kontseptualnye osnovy i prakticheskie vyvody*, 98–109.
31. Quoted in Jacob W. Kipp, "The Labor of Sisyphus: Forecasting the Revolution in Military affairs During Russia's Time of Troubles," in *Toward a Revolution in Military Affairs? Defense and Security at the Dawn of the Twenty-First Century*, eds. Thierry Gongora and Harald von Riekhoff (Westport: Greenwood Press, 2000), 91.
32. Vladimir I. Slipchenko, "For What Kind of War Must Russia Be Prepared?" in *Future War*, eds. Makhmut Gareev and Vladimir I. Slipchenko (Moscow: Ob'edinennoye Gumanitarnoye Izdatelstvo, 2005), 12.
33. His most famous work, discussed in chapter 2, developed the concept of "sixth-generation" warfare, which he defined as "non-contact" war, arguing that war had not vanished with the end of the Cold War but had been transformed by the shift from industrial to information societies.
34. Gerasimov, "Tsennost' nauki v predvidenii."
35. Sergei G. Chekinov and Sergei A. Bogdanov, "Voennaya futurologiya: zarozhdeniye, razvitie, rol i mesto v sisteme voennoi nauki," *Voennaya mysl'* no. 8 (August 2014): 19–29.
36. Ivan Vorobyev, Vyacheslav V. Kruglov, and Aleksandr Suptel, *Voennaya futurologiya* (n.p.: Obshchevoiskovaya akademiya VS RF, 2001).
37. Ivan Vorobyev, "Eshchye raz o voennoi futurologii," *Voennaya mysl'* no. 5 (May 2008): 62–67.

38. See, for example, S. A. Tyushkevich, "Neobkhodimoye uslovie razvitiya voennoi nauki," *Voennaya mysl'* no. 3 (May 2000); Valery A. Kiselyev and Ivan Vorobyev, "Istoriya i filosofiya voennoi nauki," *Voennaya mysl'* no. 2, (February 2007): 57–68.

39. S. A. Tyushkevich and Vyacheslav V. Kruglov, "Voennaya nauka: razmyishleniya o yeyo soderzhanii i razvitii," *Voennaya mysl'* no. 10 (2010): 63–68.

40. See, for example, Vladimir V. Babich, "O voennoi naukye i voine," *Voennaya mysl'* no. 12 (December 2009): 60–66.

41. Danilenko, "Ot prikladnoi voennoi nauki – k sistemoi nauke o voine," 26.

42. Mikhail P. Stepshin and Andrei N. Anikonov, "Razvitiye vooruzheniya, voennoi i spetsial'noi tekhniki i ikh vliyaniye na kharakter budushchikh voin," *Voennaya Mysl* no. 12 (December 2021): 35–43.

43. Makhonin has analyzed the Soviet and Russian approaches to military science and forecasting, concluding that the definitions and structures are broadly the same, although the post-Soviet Russian approach appears "better considered and logical." Makhonin, "K voprosu o voennoi nauki i yee obyekte," 30.

44. Gareev, "O vyirabotke u ofitserov kachestv i navyikov, neobkhodimyikh dlya proyavleniya vyisokogo urovnya voennogo iskusstva,"65.

45. Sergei G. Chekinov and Sergei A. Bogdanov, "Evolyutsiyasushchnosti i soderzhaniya ponyatiya voina v XXI stoletii," *Voennaya mysl'* no. 1 (2017): 30–43.

46. Viktor Khudoleev, "Generator proryvnyikh idei i predlozhenii," *Krasnaya Zvezda*, January 24, 2020, http://redstar.ru/generator-proryvnyh-idej-i-predlozhenij/.

47. Viktor Khudoleev, "Impuls k razvitiyu voennoi nauki," *Krasnaya Zvezda*, January 22, 2021, http://redstar.ru/impuls-k-razvitiyu-voennoj-nauki/.

48. Makhmut A. Gareev, *If War Comes Tomorrow? The Contours of Future Armed Conflict* (Abingdon: Routledge, 1998), vii–viii.

49. Valery V. Gerasimov, "Mir na granyakh voinyi," *Voenno-promyshlenniy Kuryer*, March 13, 2017, https://vpk-news.ru/articles/35591.

50. In Russia, military strategy includes assessment of the conditions and nature of future war; it is not focused solely on ends, ways, and means.

51. Valery V. Gerasimov, "Razvitie voennoi strategii v sovremmenykh usloviyakh. Zadachi voennoi nauki," *Vestnik Akademii voennyikh nauk* 67, no. 2 (2019): 7–9.

52. Tor Bukkvoll, "Iron Cannot Fight – The Role of Technology in Current Russian Military Theory," *Journal of Strategic Studies* 34, no. 5 (2011): 681–706, DOI: 10.1080/01402390.2011.601094.

53. Gareev, *If War Comes Tomorrow?*, viii.

54. The President of the Russian Federation, "Zasedaniye diskussionnogo kluba 'Valdai,'" October 21, 2021, http://kremlin.ru/events/president/news/66975.

55. Boris Shaposhnikov, *Mozg Armii*, vol. 3 (Moscow-Leningrad: Voennyi Vestnik, 1929), 239. Cited in Garthoff, *How Russia Makes War*, 11.

56. Garthoff, *How Russia Makes War*, 11. Nevertheless, war was not a goal: the Soviets preferred to achieve their objectives through nonviolent means.

57. Sokolovsky ed., *Military Strategy*, 167.

58. Y. Klein, "The sources of Soviet strategic culture," *The Journal of Soviet Military Studies* 4, no. 2 (1989): 455. Lenin believed that war was a "universal test of the material and spiritual resources" of a nation and that victory was achieved by those "endowed with the better morale."

59. Garthoff, *How Russia Makes War*, 9.

60. Sokolovsky, ed., *Military Strategy*, 11.

61. I. V. Bocharnikov, S. V .Lemshev, and G. V. Lyutkene, *Sovremennyie kontseptsii voin i praktika voennogo stroitelstva* (Moscow: Ekon-Inform, 2013), 26–27.

62. Aleksandr A. Svechin, *Strategy* (Minneapolis: East View Publications, 1992), 83–84.

63. Sokolovsky, ed., *Military Strategy*, 170.

64. Vladimirov, *Osnovyi obshchei teorii voinyi v 3 chastyakh. Chast 1: osnovyi teorii voinyi*, 53.

65. Gerasimov, "Mir na granyakh voinyi."

66. Ministry of Defense of the Russian Federation, "Voina," *Voennyyi entsiklopedicheskii slovar'*, https://encyclopedia.mil.ru/encyclopedia/dictionary.htm.

67. He does not deny the influence of other forms of struggle (*bor'ba*), such as economic, ideological, psychological, and informational, but argues that they acquire a completely different character, more violent and destructive. Gareev cited in I. V. Bocharnikov, S. V .Lemshev, and G. V. Lyutkene, *Sovremennyie kontseptsii voin i praktika voennogo stroitelstva*, 72.

68. Makhmut A. Gareev, "Voina i voennaya nauka na sovremennoi etape," in *Evolyutsiya form, metodov i instrumentov protivoborstva v sovremmenykh konfliktakh*, ed. I. Bocharnikov, (Moscow: Econ-Inform, 2015), 4.
69. Makhonin, "K voprosu o voennoi nauki i yee obyekte," 26–36. Vladimir V. Babich, "O novom podkhode k analizu sovremennogo protivoborstva i nekotorykh drugikh problemakh," *Voennaya mysl'* no. 3 (March 2008): 34–35.
70. Mikhail M. Kurochko, "Neklassicheskiye voinyi sovremennoi epokhi: k postanovke problemyi," *Elektronnii nauchnii zhurnal problemyi bezopasnosti* no. 3 (2008): 1–2.
71. Aleksandr Bartosh, "Gibridnaya voina stanovitsya novoi formoi mezhgosudarstvennogo protivoborstva," *Nezavisimoe Voennoe Obozrenie*, April 7, 2017, https://nvo.ng.ru/concepts/2017-04-07/1_943_gibryd.html/ ; Bartosh, "Agressiya novogo tipa," *Nezavisimoe Voennoe Obozrenie*, May 18, 2018, https://nvo.ng.ru/vision/2018-05-18/1_996_agression.html?.
72. Viktor N. Gorbunov and Sergei A. Bogdanov, "O kharaktere vooruzhennoi bor'byi v XXI veke," *Voennaya mysl'* no. 3 (March 2009): 2–15.
73. Tyushkevich and Kruglov, "Voennaya nauka: razmyishleniya o yeyo soderzhanii i razvitii," 66.
74. Mikhail Borchev quoted in I. V. Bocharnikov, S. V .Lemshev, and G. V. Lyutkene, *Sovremennyie kontseptsii voin i praktika voennogo stroitelstva*, 73.
75. Quoted in V. Bocharnikov, S. V .Lemshev, and G. V. Lyutkene, 74.
76. Popov and Khamzatov, *Voina budushchego*, 165.
77. Gerasimov "Mir na granyakh voinyi."
78. For example, V. V. Zhikharskii, "K voprosu o voinakh budushchego" *Voennaya mysl'* no. 4 (July–August 2000), 77–79; Makhmut A. Gareev, "O nekotoryikh kharakternyikh chertakh voin budushchego," *Voennaya mysl'* no. 6 (June 2006): 52–59.
79. I. V. Bocharnikov, S. V .Lemshev, and G. V. Lyutkene, *Sovremennyie kontseptsii voin i praktika voennogo stroitelstva*, 78.
80. V. L. Chengaev and S. V. Balenko, "Usloviya vozniknoveniya vooruzhennyikh konfliktov v XXI veke na territorii Rossiiskoi Federatsii i vozmozhnyi ikh kharakter v period obostreniya voenno-politicheskoi obstanovki," *Voennaya mysl'* no. 9 (September 2009): 2-7; S. Aliyev, "Voennaya bezopasnost' Rossii i sotsial'nyie konfliktyi," *Voennaya mysl'* no. 4, (April 2010): 3–7.

81. Vladimir I. Lutovinov, "Razvitie i ispolzovanie nevoennykh mer dlya ukrepleniya voennoi bezopasnosti Rossiiskoi Federatsii," *Voennaya mysl'* no. 5 (May 2009): 2–12.

82. Gorbunov and Bogdanov, "O kharaktere vooruzhennoi bor'byi v XXI veke," 2–15.

83. Sergei Chekinov and Sergei Bogdanov, "O kharaktere i soderzhivanii voinyi novogo pokoleniya," *Voennaya mysl'* no. 10 (October 2013): 13–24.

84. Vladimir M. Moiseev, "Oruzhiye neletal'nogo deistviya kak per-spektivnoye sredstvo voenno-silovogo vozdeisviya (kompleksnogo porazheniya protivnika)," *Voennaya mysl'* no.11 (November 2021) 41–48.

85. Kruglov and Shubin, "O vozrastayushchem znachenii uprezhdeniya pro-tivnika v deistviyakh," 27–34.

86. Gerasimov, "Razvitie voennoi strategii v sovremmenykh usloviyakh. Zadachi voennoi nauki," 10.

87. Gerasimov, "Po opytu Sirii," *Voenno-promyshlennii kurer,* March, 7, 2016, https://vpk-news.ru/articles/29579.

88. Quoted in Oleg Vladykin, "Voennaya nauka smotrit v budushchee," *Krasnaya Zvezda,* March 26, 2018, http://redstar.ru/voennaya-nauka-smotrit-v-budushhee/?attempt=2.

89. Aleksandr V. Dvornikov, "Formyi boevogo primeneniya i organizatsiya upravleniya integrirovannyimi gruppirovkami vooruzhennyikh sil na teatre voennyikh deistvii," *Vestnik Akademii voennyikh nauk* 68, no. 2 (2018): 38.

90. See for example Zarudnitsky, "Kharakter i soderzhaniye voennyikh kon-fliktov v sovremennykh usloviyakh i obozrimoi perspective," 34–44.

91. Stepshin and Anikonov, "Razvitiye vooruzheniya, voennoi i spetsial'noi tekhniki i ikh vliyaniye na kharakter budushchikh voin."

92. P. A. Dulnev, S. A. Sychev, and V. A. Garvardt, "Osnovnyie napravleniya razvitiya taktiki Sukhoputnyikh voisk (po opyitu vooruzhennogo kon-flikta v Nagornom Karabakhe)," *Voennaya mysl'* no. 11 (November 2021): 49–62; A. V. Shubin, I. V. Kot, and A. A. Kuzenkin, "Izmeneniye kontsep-tualnyikh podkhodov k primeneniyu aviatsii v voinakh budushchego na primere karabakhskogo konflikt," *Voennaya mysl'* no. 9 (September 2021): 43–50.

93. For example, see Aleksandr V. Khomutov, "O protivodeistvii protivniku v usloviyakh vedeniya im 'mnogosfernyikh operatsii'," *Voennaya Mysl'* no. 5 (May 2021): 27–42.

94. Ya A. Chizhevsky, "Osnovnyie tendentsii transformatsii prirody i kharaktera sovremennykh voenno-politicheskikh konfliktov," *Voennaya mysl'* no. 6 (June 2020): 7–8.
95. See for example the NATO definition: "Fundamental principles by which the military forces guide their actions in support of objectives. It is authoritative but requires judgement in application." The official NATO Terminology database, https://nso.nato.int/natoterm/Web.mvc.
96. Sokolovsky, ed., *Military Strategy*, 42.
97. "The Military Doctrine of the Russian Federation," approved by the President of the Russian Federation on December 25, 2014, no. Pr-2976.
98. "The Military Doctrine of the Russian Federation," approved by the President of the Russian Federation on December 25, 2014
99. Gerasimov, "Tsennost' nauki v predvidenii."

CHAPTER 2

OBSERVING WESTERN INTERVENTIONS

The systematic, scientific analysis of military trends and developments around the world is central to Russian military thought, part of enduring efforts to understand the characteristics of conflict and forecast the future character of war. International events influence Russian military thinking, and there has been frequent in-depth analysis of Western interventions such as the 1991 Gulf War (Operation Desert Storm), Operation Allied Force, and the coalition interventions into Iraq in 2003 and Libya in 2011. Military theorists have argued that, in order to foresee changes in the character of conflict, the experience accumulated by the armed forces of other countries (as well as its own operational experience) needs to be analyzed.[1] Western operations are frequently referred to in the Russian theoretical military literature and used as important examples of the way the nature of conflict is changing in the twenty-first century. Thus, these interventions have played a key role in the evolution of both Russian military thought and the country's defense policy, a fact that has prompted Russian military theorists to consider the character of conflict and war and whether it has changed. The technological advances demonstrated by the US military in its post–Cold War operations, combined with Russia's own experiences from the wars in Chechnya and Georgia, reinforced Russian concerns that its own military was lagging behind and that modernization was imperative.

This chapter explores Russian analyses of Western interventions in the post–Cold War era and the lessons drawn from them. There is widespread scrutiny of Western interventions of the twenty-first century, which has sparked among military thinkers a broad discussion about the changing character of conflict and varied ways that actors now seek to achieve their strategic objectives.[2] Aleksandr Raskin noted that one of the principal conclusions of foreign military theorists is that there is "no point in waging wars in the twenty-first century using the forms and methods from the last century and, in particular, in World War Two."[3] Operation Desert Storm and Operation Allied Force were believed to constitute an important stage in the development of a new model of war, leading to the development of forward-looking concepts such as Slipchenko's "sixth-generation warfare," based on extensive analysis of these two interventions, and "new-generation war," defined by Sergei G. Chekinov and Sergei A. Bogdanov. Slipchenko developed the concept of "sixth-generation" warfare, which he defined as "non-contact" war, arguing that war had not vanished with the end of the Cold War but had been transformed by the shift from industrial to information societies. Concepts such as asymmetric and "network-centric" wars have also been widely discussed and the implications of them for Russia analyzed. Raskin suggests that whereas previous conflicts had been initiated with the aim of routing the opposing armed forces and seizing territory, "today the aggressor is moving on to wars and armed conflicts of selective influence" while using nonmilitary means. He highlighted that in asymmetric wars, where nonmilitary means, including psychological, environmental, economic, financial, and biological, are predominant, there is no traditional front line and operations are conducted through the use of high-tech communications and the global information network.[4] Russia's invasion of Ukraine in

February 2022 has undermined these assumptions (discussed in chapter 7), highlighting the dangers of technological determinism and the myth of perpetual progress.

Analyzing Western experiences in military operations has enabled Russian military theorists to assess the relative merits of particular approaches, Russian vulnerabilities, and the suitability (or otherwise) of concepts such as network-centric warfare, which are examined in chapter 4. Observation of Western interventions has heightened concerns about Russia's vulnerabilities to emerging concepts and approaches, which has prompted Russian analysts to examine their country's defenses. Observation has also exacerbated an enduring sense of technological inferiority, particularly vis-à-vis the US. A common theme in the literature is a conviction that apparent changes in the character of conflict, such as the emphasis on precision strike and conventional weapons, has weakened the importance of nuclear weapons, the only area where Russia had strategic parity with the US. This echoes concerns raised in the early 1980s by then Soviet CGS Marshal Nikolai Ogarkov, who warned that new high-precision conventional weapons systems being developed by the US could be as effective as nuclear weapons and undermine the strategic parity between the USSR and the US.

THE 1990s: OPERATIONS DESERT STORM AND ALLIED FORCE

From a Russian perspective, the 1990s not only saw the end of the Cold War and the dissolution of the Soviet Union but also heralded the development of a new way of war, led by the US and stimulated by the emergence of the information age. Operation Desert Storm revived debates about whether an RMA was taking place or whether the changes were merely evolutionary. The use of overwhelming military power, particularly precision-strike weapons, against a less capable adversary was perceived to be both innovative and hazardous for Russia. The opening phase of the war was dominated by an extensive aerial bombardment that lasted forty-two days, targeting both military and civilian infrastructure.

This was followed by a brief ground campaign by coalition forces, who declared a ceasefire less than a week later. Desert Storm was a decisive victory for the US-led coalition, and the outcome, which surprised many within the Russian military, prompted widespread (and ongoing) debate about the character of war. Chekinov and Bogdanov have characterized the 1991 Gulf War as the first war of a new epoch, a "new-generation war" in the high-tech era. In their view, it drew a line between present warfare and the classical wars of the past, demonstrating that technological superiority can negate an adversary's quantitative advantage in obsolete conventional weaponry:

> The Iraqis demonstrated an outdated, inflexible strategy of posi-
> tional warfare, which could not withstand the new forms and
> methods of warfare used by the armed forces of the United States
> and its allies. This strategy predetermined the crushing defeat
> of the Iraqi forces. For the first time, ground forces of colossal
> proportions (half a million in number) had no impact on the
> pursuit of victory.[5]

They concluded that the US had deployed a new, uncharacteristic form of warfare, which they termed "electronic shock," a large-scale electronic operation, combined with a massive aerial bombardment in order to "disorganize" Iraqi air assets and air defense, that prevented any coun-terattack. The widespread, effective use of modern IT systems, automated command and control, and satellite technology in an integrated manner facilitated US dominance and, consequently, victory. Russian analyses of the 1991 Gulf War highlight long-running concerns about technological inferiority and the fear of being left behind by advances in military technology. Ogarkov's conclusions in the 1980s had predicted change resulting from technological advances, particularly precision-guided munitions and information technologies; events in 1991 seemed to provide evidence in support of Ogarkov's assertions and led to growing concern about simultaneous joint-strike operations and the impact of advanced communication equipment, as well as the use of information confronta-tion and subversion against an adversary. NATO's 1999 air campaign against the Federal Republic of Yugoslavia provided further empirical evidence to support this line of thinking.

Like the 1991 Gulf War, NATO's 1999 Operation Allied Force, which was led by the US, is frequently referred to in the Russian theoretical military literature, where it is used as an important example of the changing character of conflict in the twenty-first century. According to Gareev, OAF exemplifies the "modern military methods of democratic countries and their armies...aerial bombardments destroyed cities, power stations, hospitals, bridges and other key infrastructure, which forced the country's leadership to capitulate, and the ground forces were barely involved."[6] He asserted that NATO's coercive actions against Serbia marked the beginning of a new era in military conflict, one in which powerful Western organizations and states ignore the UN and international legal norms.[7] This is supported by Alexei G. Arbatov, who believes that Russia learned fundamental lessons from NATO's actions in Kosovo in March 1999: "Above all the end justifies the means. The use of force is the most efficient problem solver, if applied decisively and massively," whereas the legality of the decision to use force and humanitarian suffering are of secondary importance relative to the ends.[8]

Analysis of NATO's OAF by Russian analysts drew worrying conclusions for Russian defense. The first six weeks of remote, non-contact warfare saw the rapid destruction of Serbia's radar-based air defenses to ensure NATO's air supremacy, thereby facilitating the subsequent (and unimpeded) destruction of key economic and military infrastructure targets. Slipchenko noted that there was no theater of combat during OAF because there was no engagement of the adversary: "one side strikes from the air, the other cannot repulse the attack." He concluded that the Americans were "way ahead of all countries, including Russia: they have theatres of war, but no theatres of combat."[9] Of particular concern for Moscow was the destruction of electronic warfare and information systems, including TV, radio stations, and transmitters, prompting the inference that the US intended to ensure that the "public should not be informed about the true course of the war."[10] According to Arbatov, Russian concerns that it might find itself on the receiving end of similar precision strikes against its industry, infrastructure, and military targets prompted a new emphasis on the country's air defenses, including the development of the S-300 and S-400 surface-to-air missile systems.

Thus, NATO's air campaign against Yugoslavia was a watershed moment for Russia's relationship with the West, both in terms of Moscow's perception of the latter's intention and threat and its view of its own vulnerabilities. The anxieties it expressed in 1999 were not mere rhetoric but genuine expressions of alarm and apprehension. Russian perceptions were specifically influenced by the nature of the NATO intervention: an offensive military operation intended to coerce Serbia into accepting the alliance's terms, forcing Belgrade to capitulate and change its behavior within its own internationally recognized sovereign territory. Thus, Russian concern centered around NATO's coercive use of force against a sovereign state with the aim of changing its internal conduct, rather than its operation against Serbia per se. OAF was considered to be a potential exemplar for NATO's future approach, a deliberate move away from its traditional task of collective defense toward a strategy of unilateral military action conducted without the approval of the United Nations Security Council and with limited regard for international law, thereby undermining customary principles of state sovereignty in the name of humanitarian intervention. This prompted fears that Russia could become the victim of such an approach and therefore needed to heed the lessons from Western operations of the 1990s and early 2000s.

SLIPCHENKO AND SIXTH-GENERATION WARFARE

The 1991 Gulf War and NATO's operation against Serbia in 1999 stimulated the development of the concept of the "sixth generation" of warfare. This was formulated by Vladimir I. Slipchenko, a retired senior military officer, who taught at the Academy of the General Staff and focused on developing military science, in particular forecasting the character of future wars. He argued that war had not vanished with the end of the Cold War but had been transformed by the shift from industrial to information societies.[11] He divided the past and future into two distinct phases: the pre-nuclear, lasting 5,500 years, and the nuclear, which commenced in 1945. The pre-nuclear epoch involved predominantly contact warfare: adversaries facing each other on a battlefield. The nuclear epoch produced remote, non-contact warfare, where a country was able to strike any other country on the planet.[12] Despite being one of Bukvoll's "revolutionaries" (outlined in chapter 1), he refuted the concept of a revolution in military affairs, in which new technologies would produce a revolution, arguing instead for evolution and asserting that there have been six generations of war throughout the history of mankind (see table 1), from the first generation of warfare with edged weapons (such as spears, bow and arrows, and swords) to the sixth generation of non-contact warfare involving long-distance strikes by precision weapons.

Based on extensive analysis of the 1991 Gulf War and NATO's 1999 Operation Allied Force, Slipchenko defined sixth-generation warfare as "non-contact" war, asserting that it differed radically from previous generations of warfare because the principal objective was the adversary's economy, not territorial gain. Both campaigns commenced with intensive aerial and naval bombardments, with ground forces only being deployed during the final phase of the conflict. According to Slipchenko, the 1991 Gulf War was the first remote, non-contact war conducted by the US.

The main characteristics of sixth-generation warfare are the decisive role of high-precision conventional weapons (both offensive and defensive), along with weapons based on new physical principles; electronic warfare and the secondary role of land forces also comprise important aspects of it. Slipchenko labeled this type of warfare "contactless" or "non-contact" and stated that the principal aim of this type of operation is the destruction of an adversary's economic potential, along with a change of political regime.[13] This reflects NATO's air campaign against Serbia in 1999, which saw the widespread use of precision-guided munitions to destroy the country's economy.

Table 2. Slipchenko's Generations of Warfare.

	Characteristics					
	Weapons	Land	Maritime	Air	Principal Objective of War	Scale
First	Steel arms, no firearms	Hand-to-hand fighting	Boarding battles of the galley fleet in the coastal zone	---.	Destroy the enemy, take possession of his weapons, valuables	Tactical
Second	Gunpowder and smoothbore weapons	Frontal fire	Naval battles of the sailing fleet in the coastal seas	---.	Destroy the enemy, take possession of his territory, values	Operational-tactical
Third	Rifled small arms and tube artillery of increased rate of fire, accuracy and firing range	Trench, trench wars of combined arms formations and formations	Naval battles of steam metal ships of various classes	---.	The defeat of the enemy's armed forces, the destruction of his economy and the seizure of territory	Operational-strategic
Fourth	Automatic and jet weapons, mechanized troops, tanks, aviation, aircraft carriers, submarines	Front operations	Marine operations	Air strikes against troops, air battles	The defeat of the enemy's armed forces, the destruction of his economic potential, the overthrow of the political system	Strategic
Fifth	Nuclear missiles	Nuclear strike	Nuclear strike	Nuclear strike	It does not achieve any goals—the side that used nuclear weapons first dies a little later than the second	Strategic, global, the threat of the death of civilizations or individual continents
Sixth	Precision weapons, weapons based on new physical principles, information weapons, forces and means of electronic warfare	Joint air-sea and ground operations	Joint air-land and sea operations, aerospace and sea operations	Air operation using conventional weapons	Undermining the economy, management system, the life of the state and the destruction of military facilities	Operational-strategic

Slipchenko argued that decisive military operations would take place in the air and space domains, meaning that a country's sovereignty could be violated without the physical presence of enemy forces on its territory. In his view, there would be no need to have battlefield clashes in the future: the game-changer would be the use of information and of long-range missiles to target opponents' economic centers and their command-and-control systems (starting with government). The concept of sixth-generation warfare was based on the assumption that high-intensity war would involve the targeted destruction of key assets such as critical infrastructure using high-precision weapons rather than massed armies. Thus, it emphasizes the decisive role of high-precision conventional weapons (both offensive and defensive) and the secondary role of land forces, with the principal objective being the destruction of an adversary's economic potential and a change of political regime, rather than the seizure and holding of territory. In this type of war, conventional precision-strike weapons and information confrontation are deemed to play a decisive role, rather than a country's nuclear arsenal. Slipchenko wrote that superiority over an opponent was only possible after superiority in information, mobility, and rapidity of reaction had been achieved. From this perspective, information confrontation would have to be continuous.

In a 2001 article to mark the second anniversary of NATO's bombing of Serbia, Slipchenko asserted that, whereas the 1991 Iraq War had been a "prototype" of non-contact or contactless war, OAF constituted a comprehensive application of the method.[14] Long-range strikes by the US and its allies targeted critical military, economic, transport, and communications infrastructure, transferring war to the air domain, which became, according to Slipchenko, the "principal theatre of war." This meant that space assets were considered to be playing a "critical, systemic" role, along with electronic warfare, the latter of which Slipchenko believes was used to suppress Serbia's air defense systems. He concluded by stating that OAF had facilitated the testing of the Untied States' global command system and had demonstrated a "hidden arms race for wars of the future," in which precision-guided munitions were the principal and long-term source of income for many defense industrial companies.[15] Speaking in 2004, Slipchenko once more emphasized his belief that the US and its NATO allies were moving toward a new generation of warfare, the "remote, non-contact generation," which would be the warfare of the future:

> Those are the types of wars for which Russia must prepare. No-one is ever going to come to us by land again. These days it is impossible to imagine armoured spearheads crashing across the western, southern and eastern borders. If war reaches us it will reach us via aerospace and the strike will come from precision weapons.[16]

He concluded that Russia was still in the fourth generation of warfare; according to his analysis, Russia was focused on the contact warfare of era of the Great Patriotic War, whereas the US had been conducting non-contact warfare for over a decade. This reflects long-running concerns among military theorists and policy makers that Russia was lagging behind strategic competitors and was not preparing for the "right" type of war. Gareev (who had worked with Slipchenko on issues surrounding future war) built on these ideas in a 2006 article, arguing that precision-guided long-range weapons would change the character of future conflict. He warned that Russia should not underestimate the impact of such weapons, asserting that, in the future, "wars will be contactless," with no soldiers on the ground to take prisoner: "the principal...feature of wars and armed conflict of the future will be the exclusion of people from the area of confrontation."[17]

Slipchenko's work is not without its detractors. There was criticism from some Russian military theorists of his methodological approach, which was condemned as weak and leading the author to a number of "highly controversial conclusions."[18] Using the example of nuclear weapons, Vasily G. Reznichenko argued that throughout history countermeasures have quickly been developed to address the threat of new weapons. He accused Slipchenko of "isolation from reality" and urged him to adopt a more scientific approach to the study of future conflict, noting that for many years the protection of troops and critical state facilities from high-precision strikes have been analyzed by both the government and research institutions within the MoD.[19] Slipchenko's work was also criticized for not setting out a clear definition of "contact-less war" or its essence.[20] Valentin Rog condemned Slipchenko's thesis, in particular his assertion that contactless war renders air defense forces irrelevant, pointing to the US missile defense program as evidence of its continued utility.[21]

Slipchenko's second book, *Contactless War*, published in 2001, built on his previous work and introduced the concept of a "formula of victory" in sixth-generation warfare, asserting that, on the basis of analysis of a wide variety of conflicts, three objectives had to be attained in order to achieve victory: the first is to "defeat the enemy's armed forces, as a rule, on his territory, the second is to destroy the economic potential of the enemy, [and] the third is to overthrow or replace the political system of the enemy. If these three components were achieved simultaneously, the victory was considered complete. But if one of the three areas of success has not been achieved (as was the case in the Persian Gulf in 1991: Hussein remained in power), the victory cannot be complete."[22] According to this analysis, comprehensive victory is no longer achieved by the occupation of territory but through the destruction of an adversary's economic potential and the response of the local population, who may well seek to overthrow the existing regime. Slipchenko claimed that the most developed countries were transitioning toward sixth-generation, "contactless," wars, and those countries left behind were preparing for fourth-generation contact war and fifth-generation contactless nuclear war: consequently, the political elites in less developed countries were betting on nuclear deterrence as a means of ensuring military security.

This focus on the central role of nuclear deterrence in countries that did not have long-range precision strike capabilities accounts, to some extent, for Moscow's opposition to the US missile defense program, which was considered to pose a significant threat to Russia. The US announcement in 2001 that it intended to withdraw from the Anti-Ballistic Missile Treaty laid the groundwork for continuing tensions between Moscow and Washington. Withdrawal from the treaty—the first time the US had ever withdrawn from a major international arms control pact—enabled it to proceed with plans to construct a missile defense system (initially with a purely national focus). Sergei Rogov argues that Washington's actions demonstrated that the US was no longer interested in the appearance of strategic parity and had embarked on a course intended to ensure absolute military superiority, with no intention of recognizing any other state's "equal strategic status." Consequently, Russian theorists perceived US plans for a national missile defense system to be a significant strategic threat to Russia, undermining the one area in which it retained parity with the US. Although the initial focus of the planned missile defense was on homeland defense, a European pillar was unveiled in 2006 with the announcement of plans to install interceptors in Poland and a radar control center in the Czech Republic. Moscow rejected US assurances that the expansion of the missile defense system posed no threat to Russian forces, viewing the move as a deliberate attempt to consolidate its global dominance and intimidate potential challengers.[23] Slipchenko accused the US of seeking to develop a national missile defense system as part of its move toward contactless war, arguing that the US needed a missile defense umbrella "like a fish needs an umbrella."[24] Underpinning this criticism was a long-running concern about Russia's technological inferiority and the gap between the two states in all fields other than strategic nuclear weapons.

IRAQ 2003

Debates about the central role of precision-strike in Western interventions and the ramifications for Russian defense were renewed in analyses of US-led coalition operations in Iraq, which began in March 2003. There were a number of articles assessing operations and the lessons that Russia should and could draw from Western military experience, and the Academy of Military Science also held a series of discussions to analyze the actions of the coalition forces in Iraq.[25] Building on Slipchenko's work on sixth-generation warfare, theorists continued to focus on the use of high-precision weapons and contactless warfare, with one analysis concluding:

> The strongest aspect of the American army is the provision of long-range precision weapons to all branches of the armed forces... facilitating non-contact combat operations. In addition to large losses...such superiority undermines the [adversary's] will to resist.[26]

One of the principal lessons that Russian analysts drew was the key role of air superiority in facilitating US dominance. In an assessment of the role of US air power in Iraq, V. Zayats writes that advanced technology, particularly aircraft and precision-guided weapons, had become critical to victory in modern wars.[27] Anatoly D. Tsyganok , head of the Center for Military Forecasting, argues that the US (and the West more broadly) overestimated the capabilities and impact of high-precision weapons to act as the decisive factor in military operations, noting that since the 1991 Gulf War—and subsequently in Yugoslavia and Afghanistan— the decisive factor was the "political isolation of leaders" rather than military superiority.[28] Interestingly, the overwhelming dominance of coalition forces was perceived to have undermined the role of strategic surprise. An early Russian analysis of the 2003 intervention in Iraq identified a number of distinctive features of the operation, including a high level of interaction between different services/branches of the armed forces. There were joint and combined interactions; the intensive and simultaneous use of "vast" quantities of modern weapons and military equipment, necessitating high-quality command and control; the large-scale use of guided weapons systems, based on satellite data systems;

and the criticality of achieving information superiority prior to any military action.[29]

The information domain was considered to have become a vital component of conflict. A number of authors noted the importance placed on information support in US foreign policy initiatives, particularly the use of force overseas, and suggested that the US was engaged in information and psychological operations long before the direct invasion of Iraq, with the aim of creating favorable conditions for the removal of Saddam Hussein.[30] The US was perceived to have been waging a covert operation of nonmilitary means against Iraq before the military phase of the campaign began in March 2003, with the aim of undermining the country's "will to resist," demonizing the Iraqi leader and emphasizing the alleged threat posed by weapons of mass destruction:

> Before starting a real war (military operations), an economic blockade, political isolation, targeted information operations on the population are conducted for a long time, international public opinion is shaped...., and measures are taken to weaken military potential...In general, such powerful political, economic and psychological pressure was exerted on the people and armed forces that they could not think of any kind of resistance. And then missile strikes rained down to finally break the will to resist.[31]

Based on the evidence of US operations in Iraq, information warfare operations were perceived to be divided into three stages. Prior to the deployment of force, a PR campaign would be conducted to gain domestic support for an intervention and demonize the adversary. Once armed conflict commenced, information warfare would switch to the operational-tactical level with the aim of persuading and coercing the adversary into an unconditional surrender through information and propaganda actions, such as media broadcasts and leafleting via airdrop, as well as electronic warfare. The final stage is the "interpretation" of events to benefit the aggressor.[32] The contemporary era is characterized an "era of information warfare" in which the role of space-based assets is critical to military operations. An absence of information is seen as depriving decision makers of direction and leads to "unjustified subjective decisions," meaning that information is as much a military resource as soldiers and military equipment.[33]

Specific battles from coalition operations in Iraq were also analyzed, including the battle for Mosul. Valery A. Kiselyev and Alexei N. Kostenko argue that the US struggle for control of Mosul challenged the widespread belief in the overwhelming superiority of US troops. In their view, it also demonstrated that key aspects of the Western approach, including a reliance on high-precisions munitions, aviation, air defense, and network-centric warfare, were ineffective in counterterrorist and counterinsurgency operations.[34] Common lessons identified in a range of Russian analyses include the decisive role of air and information superiority, facilitated by effective command and control, as well as the importance of space-based assets. Analysts perceived Russia to be lagging behind Western forces due to its limited reconnaissance, communications, and logistics capabilities, and there were calls for the country to focus on the creation of more advanced electronic warfare systems.

Some military theorists have sounded a note of caution, concluding that it is impossible to draw any far-reaching conclusions about the development of military art or the character of conflict from coalition operations against Iraq because it was not a "serious war against a serious adversary." Instead, the campaign was deemed to be a retaliatory attack by a "politically sophisticated and technologically powerful state" over an obviously weakened adversary.[35] Nevertheless, there is still a firm belief that events in Iraq underscored the significance of the foundations of Russian military science and military art.[36] The Russian Academy of Military Science asserted that the war in Iraq accentuated the enduring relevance of ground forces, refuting Slipchenko's thesis that contemporary conflicts would be "contactless." It also warned that Russia should not follow the example of other states and focus purely on counterterrorism, arguing that the country needed a powerful, diversified Armed Forces capable of operating across the spectrum of conflict.[37]

LESSONS FROM THE WEST

There has been a substantial focus on the lessons that Russia can draw from Western military operations (particularly those of the US and NATO) over the past decade. A good example of this is a 2009 article by Gorbunov and Bogdanov, which analyzes the character of armed conflict in the twenty-first century, drawing upon examples from Western interventions to illustrate the unpredictable nature of contemporary wars and conflicts, as well as new forms of warfare wherein military means either lack a stable role or do not have a role at all.[38] They argued that without an "in-depth analysis of the relationship between peace and war in the twenty-first century, it is impossible to make any logical prediction of future war." Noting that changes in the international system, as well as global and regional security structures, were ongoing and unpredictable, they sought to classify contemporary wars and conflicts, setting out what they perceived to be the most important trends in the character of conflict in the twenty-first century:

- An increasing use of weapons based on AI and nanotechnology, as well as both weapons and systems based on "new physical principles";
- An expectation that the capabilities of these weapons will increase further, potentially given them the power and effectiveness of nuclear weapons;
- An increasing role for and importance of forces and assets operating in the air domain;
- The increasing importance of informational components of armed conflict as a result of advanced communications equipment and networked weapons systems, meaning that the pursuit of information superiority over an adversary will be vital for success in future operations;

- A change in the temporal parameters of armed conflict, including the compression of mobilization and military operations;
- A transition from a vertical hierarchy of command toward a global networked command and control system that is increasingly automated;
- And the increasing role of special forces, both for reconnaissance and sabotage, as well as supporting opposition groups to organize armed resistance and insurgencies.

On the basis of their analysis of the wars and conflicts of the post–Cold War era, Gorbunov and Bogdanov conclude that future conflict would be multi-domain and facilitated by technological advances which would both intensify the impact of a military operation and compress its duration, enabling an actor to achieve "decisive results in the shortest possible time" and deprive an adversary of the initiative and freedom of maneuver.[39] The authors argue that the weakening of a state in order to deprive it of the will to resist is a key objective of contemporary aggression, an objective that could be achieved by undermining the state's external position through a variety of methods, including surrounding it with allies of its adversary. They use the example of OAF in 1999 to illustrate their point. One of the key themes of their work is an analysis of developments in US policy and strategic thinking, as well as the contemporary operational experience of the US and NATO, and their analysis leads them to conclude that Russia still lagged behind in military terms.

Citing Svechin's famous quote that each war is particular and that there is no template, Bogdanov noted in 2012 that almost all of the armed conflicts of the twenty-first century had differed from each other, both in terms of content, the actors involved, duration, and forms and methods.[40] Nevertheless, through an analysis of NATO's action against Yugoslavia in 1999 (and comparing with it with the US's intervention in Iraq in 1991), Operation Enduring Freedom in Afghanistan from 2001, and Iraq in 2003, he identified what he argued were common characteristics:

- Planning was led by the US and, a few months before the start of a conflict, the military-political leadership sought to create "favorable" conditions for the use of force and large-scale measures, including information, psychological, ideological, diplomatic, and economic confrontation;
- Information superiority over an adversary was key. The US was deemed to have used all forms of media to make the case for the necessity of conflict and also to undermine the target state's will to resist;
- Direct military action was preceded by reconnaissance activities to identify critical state and military facilities;
- Massive air strikes opened the military phase of a campaign, whereas ground troops were deployed only after the destruction of critical military and governmental targets.

Bogdanov concluded by stating that significant changes were anticipated in the nature of future wars and warning that Russia was not yet prepared to repel such aggression, with industries such as electronics, communications, and computer technology (all viewed as critical in contemporary conflicts) in crisis, while the level of weapons and equipment classed as "modern" within the Russian Armed Forces at that time was estimated to be less than thirty-nine percent.[41] He set out "basic forms and methods" for how Russia should seek to counter threats to its own national security, including the deployment of a strong Russian defensive group; the "unrestricted" use of high-precision weapons to target nuclear power plants, chemical industry facilities, and other industrial targets; and the use of a large-scale information campaign, both to give false information to the adversary about intended Russian action and to underscore the "negative motives and intentions" of the enemy: "on the basis of the experience of past military conflicts, it is necessary to clearly understand that all media must be manageable."[42]

Bogdanov went on to develop some of these ideas further in a 2013 article he coauthored with Chekinov, in which they set out their concept of "new-generation war" (which they first mentioned in a 2012 article), discussed in chapter 1. The concept was based on an analysis of conflicts and wars that had taken place since the end of the Cold War, combined with the views of experts on the likely characteristics of future war. In their view, technological advances had already led to a change in the forms and methods of war and would have a decisive influence in future.[43] They emphasize the high importance of asymmetric actions aimed at neutralizing the enemy's military superiority through the combined use of political, economic, technological, ecological, and information campaigns, stressing that "decisive battles will rage in the information environment":

> In a new generation war, the leading role will be played by infor-mation and psychological warfare aimed at achieving superiority in terms of the command and control of troops and weapons, and the moral and psychological suppression of the adversary's armed forces personnel and population. It is the information-psycho-logical struggle...that will largely create the prerequisites for achieving victory.[44]

Information superiority and the principle of "the first to see, the first to act decisively"—that is, holding the strategic initiative—are identified as critical components of success in new-generation wars. Nevertheless, despite this focus on the nonmilitary means, they also emphasized military means, stating that the overwhelming military technical superiority of one side would be a characteristic feature. They investigated the concept of network-centric warfare, which they defined as a concept of command and control, reflecting a new way of commanding armed forces in the twenty-first century. In their view, network-centric warfare had evolved as a result of the rapid development of IT and the creation of high-precision weapons and weapons based on new physical principles.[45] Their views on the military phase of a conflict reflected Slipchenko's sixth-generation warfare, likely to be characterized by massive aerial bombardment targeting an adversary's critical military and economic infrastructure with the aim of forcing capitulation.

The two outlined two phases of war: the opening phase, which is the longest, and the closing phase. In their view, the opening phase begins with an intensive, months-long coordinated nonmilitary campaign against the target country, including diplomatic, economic, ideological, psychological and information means. The closing phase, when military forces are deployed, is much shorter as military force is only used once a state's will to resist has been undermined during the opening phase. Chekinov and Bogdanov consider the first, predominantly nonmilitary phase of the conflict to be much more important than the second. The principal aim of the first phase is to reduce the need to use miltary force to the minimum necessary by undermining the adversary's will to resist. However, military force is still seen as retaining a significant role in enabling the realization of strategic objectives, alongside indirect, non-contact operations.

The term new-generation war has not subsequently been developed by other writers, but many of the themes in Chekinov and Bogdanov's article have been, particularly their focus on nonmilitary means. Similar to Chekinov and Bogdanov, Valentin Runov and Sergei Rodionov also published an analysis of the characteristics of contemporary conflicts in 2013 based on predominantly Western interventions, including the 1991 Gulf War; OAF in 1999; actions in Afghanistan, Iraq, and Libya; and Russia's own conflict with Georgia in 2008.[46] Noting that each conflict was distinct in terms of the use of forces and means, they also identified a number of common features. In their view, the air campaign constituted the most important component of contemporary war, preceding any ground campaign and targeting critical infrastructure as well as command-and-control units of an adversary's armed forces. Electronic warfare and reconnaissance were essential aspects of the air campaign, the success of which was sometimes consolidated with a joint air-land operation. The deployment of special forces was another characteristic, particularly in conflicts involving non-state actors such as terrorist groups. Their conclusions echoed the work of Slipchenko and others, emphasizing the centrality of precision-guided munitions and long-range strike.

The Russian General Staff and its supporting military research structures have noted the rapidly changing character of war. In 2017 Valery V. Gerasimov explored the transformation of contemporary wars and conflicts, referencing the recent operational experiences of the US and NATO. Repeating a phrase he has used frequently, he noted that there is a "blurring of the lines between war and peace" in contemporary conflicts and proclaimed that military force is increasingly used to "protect a state's economic interests under the guise of defending democracy or the promotion of democratic values" in a particular country.[47] Gerasimov noted that conflicts of the twentieth and twenty-first centuries differ from one another in various ways: by participants, weapons used, the forms and means of military activity, and the changing ways various means of struggle contribute to achieving political aims. This was echoed by the head of the General Staff's Military Academy, General Vladimir B. Zarudnitsky, who concluded, based on the examples of operations in Afghanistan, Iraq, Libya, and Syria, that in addition to long-range precision strike, military measures are frequently implemented in a "latent form," through indirect or asymmetric action, increasing the role of expeditionary, mobile units and operations with limited operations.[48] Other Russian analysts have reached similar conclusions on the basis of evidence from Western interventions in Yugoslavia, Afghanistan, and Libya: "the application of military force is preceded by a long period of political, economic, and informational pressure with a gradual escalation to military conflict."[49]

There have also been questions over the extent to which Western interventions since 1991 are indicative of change in the character of conflict. A. V. Smolvy, V. V. Loik, and K. A. Trotsenko point out that whereas wars in Iraq, Afghanistan, and Libya involved qualitatively different weapons and limited troop deployments, the lack of decisive victory (or defeat) has led to extended periods of instability both within the target countries and regionally.[50] They also cast doubt on Slipchenko's assertions that ground forces played only a secondary role in contemporary conflict, stating that they were "returning to the epicentre of armed conflict" as the only means of resolving a multilateral military conflict, although they were fundamentally different than the troops of the 1990s, with increased mobility and autonomy.

What emerges from a lot of Russian military theoretical writing is an apparent belief that the US gets involved in conflicts specifically to gain operational experience and test new weapons and military equipment. Writing in *Voennaya Mysl'* in 2017, Gareev noted that the US never misses an opportunity to give its armed forces combat experience and test new weaponry.[51] Slipchenko has also suggested that the US has launched military operations overseas in order to test weapons and systems, asserting that the "results of field experiments, obtained during the 1999 strikes on Iraq and Yugoslavia, intensified....a hidden arms race for the wars of the future."[52] He went on to state:

> The most important (if not the most important) goal of the war in Yugoslavia for the United States and its NATO allies was further comprehensive testing in real combat conditions of new, high-precision weapon systems, intelligence, command, communications, navigation, electronic warfare, and all types of support.[53]

Finally, although the focus of most Russian experts has been on Western-led interventions, there has also been increasing attention paid to the experience of Turkey (as well as China and Iran), notably with regards to the use of UAVs. The US and Israel are perceived to have lost their monopoly on the production and use of UAVs, demonstrated during the Second Nagorno-Karabakh War in 2020, which is seen as a clear example of the impact of technology on the character of conflict. Kruglov and Shubin argue that hostilities in Nagorno-Karabakh revealed how UAVs are reshaping warfare and point to Turkey's development and deployment of them in Syria and Libya and their export to Azerbaijan and Ukraine: footage of Azerbaijani UAVs attacking targets in Nagorno-Karabakh is described as "visual symbols of the celebration of a new era: the era of remotely piloted aviation."[54] The authors recognize that this era began in the early 2000s but argue that there has been a growing trend toward developing methods for using UAVs in combat. The war in Ukraine reinforced this notion: the Ukrainians made very effective use of Turkish TB2 Bayraktar UAVs to target Russian forces, the same UAVs used by Azerbaijan in Nagorno-Karabakh in 2020.

There is little question that international developments influence Russian military thinking and Western interventions since 1991 have had a significant impact on the evolution of both Russian military thought and the country's defense policy. The development of indigenous concepts such as sixth-generation warfare and new-generation war emphasize the forward-looking nature of Russian military thought, as well as the use of observation to extrapolate lessons for Russian. Observation of Western interventions has been instructive for Russia's community of military theorists, strategic thinkers, and policy makers, highlighting concerns about the country's own vulnerabilities, including disquiet about its perceived technological inferiority vis-à-vis the US and its NATO allies, as well as drawing attention to changes in the character of conflict that Russia needed to adapt to. The acquisition and use of modern high-precision weaponry by the US and its NATO allies was perceived to constitute one of the principal threats to Russian national security, and prompted a shift in procurement focus.[55] A common theme in the literature is a conviction that apparent changes in the character of conflict, such as the emphasis on precision-strike, air superiority and conventional weapons, has weakened the importance of nuclear weapons, the only area where Russia had strategic parity with the US. Analysis of US actions during operations in Afghanistan from 2001 and Iraq in 1991 and 2003 warned that Russia "needs to pay attention: we are far behind. There is a covert high precision arms race in which we are as yet lagging behind."[56] This reflects an enduring focus of Russian military thought on the role of advanced technology in future wars, which is perceived to have been critical in Western interventions such as Operation Desert Storm. The impact of technological advances on warfare will be explored further in chapter 4, as well as emulation of Western approaches: a notable aspect of Russia's intervention in Syria has been its role as a testing ground for new weapons systems, as well as an arena for troops to gain combat experience, a practice that many Russian writers have accused the US of. Russia's own experiences of conflict during the post–Cold War era, discussed in chapter 3, have also been instrumental in shaping military thought and views on the character of twenty-first-century conflict.

Notes

1. Makhonin has analyzed the Soviet and Russian approaches to military science and forecasting, concluding that the definitions and structures are broadly the same, although the post-Soviet Russian approach appears "better considered and logical." Makhonin, "K voprosu o voennoi nauki i yee obyekte," 26–36.
2. See, for example, Gorbunov and Bogdanov, "O kharaktere vooruzhennoi bor'byi v XXI veke," 2–15; Popov and Khamzatov, *Voina budushchego: kontseptualnye osnovy i prakticheskie vyvody.*
3. Aleksandr Raskin, "Gryadut li 'setevyie' sechi?," *Armeiskii sbornik* 9 (September 30, 2005): 20
4. Raskin, 20–21.
5. Chekinov and Bogdanov, "O kharaktere i soderzhivanii voinyi novogo pokoleniya."
6. Makhmut Gareev, "Problemy strategicheskogo sderzhivaniya v sovremennykh usloviyakh," *Voennaya mysl'* no. 4 (April 2009): 8.
7. Makhmut Gareev, "Rossiya dolzhna snova stat' velikoi derzhavoi," *Voenno-promyishlenniyi kur'er* 218, no. 2, (January 16–22, 2008): 10.
8. Alexei G. Arbatov, "The Transformation of Russian Military Doctrine: Lessons Learned from Kosovo and Chechnya," *The Marshall Center Papers* no. 2 (2000): v.
9. Slipchenko, "For What Kind of War Must Russia Be Prepared?," 20.
10. Gareev and Slipchenko, 21.
11. Vladimir I. Slipchenko, *Voina budyshchego: shestoye pokolenie* (Moscow: Moskovskii obshchestvennyi nauchnyi fond, 1999).
12. Slipchenko, "For What Kind of War Must Russia Be Prepared?," 4.
13. Slipchenko, *Voina budyshchego*, 27.
14. Vladimir I. Slipchenko, "Beskontaktnoye istrebleniye. Ispolnilas' vtoraya godovshchina co dnya okonchaniya agressii NATO protiv Yugoslavii – proobraza voin XXI veka," *Nezavisimoe Voennoe Obozrenie* no. 21 (June 15, 2001): 2.
15. Slipchenko, "Beskontaktnoye istrebleniye. Ispolnilas' vtoraya godovshchina co dnya okonchaniya agressii NATO protiv Yugoslavii – proobraza voin XXI veka," 2.
16. Slipchenko, "For What Kind of War Must Russia Be Prepared?," 13.

17. Gareev, "O nekotoryikh kharakternyikh chertakh voin budushchego," 52–59.
18. Vasily G. Reznichenko, "O metodologii issledovaniya voin budyshchego," *Voennaya mysl'* 1 (2003): 71–72.
19. Reznichenko, 72.
20. Valentin Rog, "Prioritet – dostizhenie gospodstva v vozdukhye. Primeneniye vyisokotochnogo oruzhiya trebuyet adekvatnykh sredstv bor'byi c nim," *Nezavisimoe Voennoe Obozrenie* (August 31, 2001): 3.
21. Rog, 3.
22. Vladimir I. Slipchenko, *Beskontaktnyie voinyi* (Moscow: Gran-press, 2001).
23. Sergei Rogov, "The Bush Doctrine," *Russian Social Science Review* no. 44, 3 (2003): 4–28.
24. Vladimir I. Slipchenko, "Aktual'no! Opyat pro?," *Armeiskii Sbornik* no. 1 (2002): 17.
25. For example, see "Uroki i vyivodyi iz voinyi v Irake," *Voennya mysl'* 7 (2003): 58–78 and "Uroki i vyivodyi iz voinyi v Irake," *Voennya mysl'* 8 (2003): 68–80.
26. "Uroki i vyivodyi iz voinyi v Irake," *Voennya mysl'* 8 (2003): 74.
27. V. Zayats, "Voenno-vozdushnyie cilyi. Primenenyie aviatsii SShA na aktivnoi faze operatsii v Irake," *Zarubezhnoe voennoe obozrenie* no. 10 (October 2005) 37–44.
28. "Uroki i vyivodyi iz voinyi v Irake," 76.
29. "Uroki i vyivodyi iz voinyi v Irake," 61–63.
30. For example, A. Saveliev, "Information Support of the Armed Forces of the US (the case of the Iraq invasion 2003)," *Zarubezhnoe voennoye obozrenie* no. 12 (December 2015): 56–62.
31. "Uroki i vyivodyi iz voinyi v Irake," 73.
32. "Uroki i vyivodyi iz voinyi v Irake," 65.
33. "Uroki i vyivodyi iz voinyi v Irake," 75.
34. Valery A. Kiselyev and Alexei N. Kostenko, "Bor'ba za Mosul v Irake kak zerkalo taktiki amerikantsev po ovladeniyu gorodami," *Voennaya mysl'* no. 2 (February 2018): 33–42.
35. "Uroki i vyivodyi iz voinyi v Irake," 73.
36. "Uroki i vyivodyi iz voinyi v Irake," 74.
37. "Uroki i vyivodyi iz voinyi v Irake," 74.
38. Gorbunov and Bogdanov, "O kharaktere vooruzhennoi bor'byi v XXI veke," 2–15.
39. Gorbunov and Bogdanov, 5–6.

40. "Each war represents a partial case, requiring the establishment of its own peculiar logic, and not the application of some sort of model" Aleksandr Svechin cited in Sergei A. Bogdanov, "Odinakovyikh voin ne byivaet," *Armeiskii sbornik* 7 (July 2012): 2–5.
41. Bogdanov, 3–5.
42. Bogdanov, 4–5.
43. Chekinov and Bogdanov, "O kharaktere i soderzhivanii voinyi novogo pokoleniya," 17.
44. Chekinov and Bogdanov, 17.
45. Chekinov and Bogdanov, 19.
46. Valentin Runov and Sergei Rodionov, "Kharakternyie chertyi sovremennoi operatsii," *Armeiskii sbornik* no. 12 (December 2013) 59–60.
47. Valery V. Gerasimov, "Sovremennyie voinyi i aktual'nyie voprosyi oboronyi stranyi," *Vestnik Akademii voennykh nauk* no. 2, 59 (2017): 11.
48. Zarudnitsky, "Kharakter i soderzhanie voennyikh konfliktov v sovremennykh usloviyakh i obozrimoi perspective."
49. A. V. Galkin, "Formyi boevogo primeneniya i organizatsiya upravleniya integrirovannyimi gruppirovkami vooruzhennyikh sil na teatre voennykh deistvii," *Vestnik Akademii voennykh nauk* no. 2, 55 (2016): 51–54.
50. Aleksandr V. Smolvy, V. V. Loik, and K. A. Trotsenko, "O nauchnoi kritike v voennom dele," *Voennaya Mysl* no. 10 (October 2021): 148–156.
51. Gareev, "O vyirabotke u ofitserov kachestv i navyikov, neobkhodimyikh dlya proyavleniya vyisokogo urovnya voennogo iskusstva," 67.
52. Slipchenko, "Beskontaktnoye istrebleniye. Ispolnilas' vtoraya godovshchina co dnya okonchaniya agressii NATO protiv Yugoslavii – proobraza voin XXI veka," *Nezavisimoe Voennoe Obozrenie* no. 21 (June 15, 2001): 2.
53. Slipchenko, 2.
54. Kruglov and Shubin, "O vozrastayushchem znachenii uprezhdeniya protivnika v deistviyakh," 46.
55. V. Litvinenko and S. Yastrebov, "VTO: vzglyad v budushchee," *Armeiskii sbornik* no. 8 (August 2017): 9–17.
56. Slipchenko, "For what kind of war must Russia be prepared?" in *Future Wars*, Gareev and Slipchenko, 23.

CHAPTER 3

OPERATIONAL EXPERIENCE IN THE POST-SOVIET ERA

The Kremlin has used the military instrument on a number of occasions since 1991 to achieve broader strategic and foreign policy goals, achieving a number of "firsts" in the process: the 1994–1996 First Chechen War was post–Soviet Russia's first war (although it had provided military support and peacekeepers for conflicts such as Abkhazia); Georgia was Russia's first war of the post-Soviet era against a foreign state; and Syria was portrayed as Russia's first Western-style "intervention," fought as much as possible at distance, either through the use of long-range precision strike or proxy forces. One of Putin's first priorities on taking power in 2000 was to halt the perceived decline of the Russian armed forces, which have undergone a comprehensive program of reform and modernization. The 2008 conflict with Georgia demonstrated the renewed ability of the Russian armed forces to fight conventional wars, following years of conflict in Chechnya and the North Caucasus. Subsequent interventions in Ukraine and Syria demonstrated the results of processes that have been ongoing in the Russian military and in Russian strategic thought over the past few decades, driven partly by Russian perceptions and understanding of the military activity of the West, along with its own operational experience since the end of the Cold War. Twenty years after its early failures in Chechnya, Russia's ambiguous use of force in Ukraine (2014–2022) was seen by many to represent significant change in the way that Moscow wields its military power, whereas its intervention in Syria revealed a qualitative change in the approach of the Russian armed forces, as well as the weapons and military equipment deployed, reflecting lessons learnt over two decades.

Russia's military interventions in Chechnya, Georgia, and Syria are commonly discussed in the military theoretical literature, although Ukraine pre-2022 is not (largely because Russia consistently denied that it was involved). This chapter explores the shortcomings identified, as well as what was perceived to have worked well, in these interventions. What lessons has Russia learned from its own operational experience? And how well have these lessons been integrated into subsequent operations? This chapter does not go into the detail of each conflict; rather, it seeks to focus on the key lessons that have been taken from them. Although the political and security elites have focused on NATO, and the West more broadly, as adversaries that Russia needs to be able to compete with in the military realm, Russian operational experience in the post-Soviet era has included combat against non-state actors and involvement in complex conflicts encompassing a variety of different actors.

CHECHNYA

The 1990s were a period of turmoil and change for Russia, politically, economically, and militarily, and the country was seen to be at its weakest at home and abroad. This was encapsulated by the first attempt to quell separatism in Chechnya when the Russian armed forces were initially unable to convert their extensive (numerically at least) military capabilities into military and strategic success, and thousands of Russian troops proved unable to secure the North Caucasian republic. The Chechen conflicts offer an interesting case study into how the Russian military adapts to lessons learned at the tactical, operational, and strategic levels; with three years between the end of the first campaign and the start of the second, the military and political leadership had time to learn lessons from the first conflict and apply them. The deleterious impact of the disintegration of the USSR in 1991 on the armed forces undermined combat readiness, and the Russian military experienced a very steep learning curve in Chechnya because the war that began in 1994 was not what they were prepared for: they had trained for high-intensity war against NATO, not low-intensity counterinsurgency. The experience in Chechnya demonstrated an apparent disregard among those in command for lessons learnt in Afghanistan,[1] meaning that the initial invasion, launched following the failed intervention by proxy forces in November 1994, was a humiliating failure. Despite the relatively recent Soviet experience in Afghanistan against an adversary using similar tactics, Russian forces were not prepared for counterinsurgency warfare against highly armed and motivated guerrillas and did not have the initiative against an adversary that had been preparing for a Russian military intervention since 1991. The Russian action lacked credibility and capability in the eyes of their adversary, as well as in the eyes of the wider Russian domestic population, undermining Moscow's resolve and credibility further. This was post-Soviet Russia's first televised war, and the federal authorities lost the propaganda war at a very early stage of the campaign.

The Russians had also been observing Western interventions, and Arbatov believes that Russia learned from NATO's actions in Kosovo in March 1999, noting that "above all the end justifies the means. The use of force is the most efficient problem solver, if applied decisively and massively."[2] Eugene Miakinkov asserts that during its second post-Soviet operation in Chechnya, Russia shifted the traditional counterinsurgency matrix to place an emphasis on the use of overwhelming superiority of force, as well as winning the support of Russian public opinion, rather than the hearts and minds of the Chechen civilian population. He identifies "devastating and almost indiscriminate firepower" as a cornerstone of the Second Chechen War (launched in 1999), unimpeded by any Russian public or political opposition, largely as a result of the demonization of the Chechens, which had taken place in the run-up to the commencement of the war in 1999, as part of Russia's information operations.[3] Having observed the crucial role of the media during NATO's Operation Allied Force, Russia took steps to control the flow of information throughout its second intervention in Chechnya. Key lessons that the Russians learned during the course of the Chechen campaigns include the difficulties of urban combat; the importance of controlling information flows; how to fight at distance using artillery, airpower, and proxy forces; and the centrality of command and control, particularly the coordination of different force structures.

Losing during the Initial Period of War

As discussed in chapter 2, Soviet-era principles that focus on time, speed, pace (rapidity), and surprise during the initial period of war continue to shape military thinking in post-Soviet Russia. Such thinking aims toward seizing the strategic initiative and achieving results in the shortest period of time by stunning the enemy, paralyzing their will, and undermining their ability to resist. However, these principles appeared to have been abandoned at the beginning of the first Chechen campaign in late 1994. During November of that year, covert Russian attempts to destabilize the regime of Dzhokhar Dudayev culminated in a failed intervention by proxy forces, which relied heavily upon Russian equipment, aircraft, and personnel. By the time the Russian armed forces (including troops from the MoD and Interior Ministry) commenced their invasion of Chechnya on December 11, 1994, pouring into the republic from different three directions, they had lost the element of surprise.[4] Chechen forces were prepared for an invasion and Russian progress was slow, hampered by civilian blockages, breakdowns, poor weather, and dissent among senior military figures who were hesitant about their involvement in what was essentially seen as a political struggle that could not be solved by military means.[5] Russian forces sought to regain the initiative with a massive armored offensive against the Chechen capital Grozny, attacking from the north, east, and west of the city on New Year's Eve. The intention was to stage a decisive strike with air support, relying on speed to take the Chechen leadership by surprise and ensure Russia held the initiative. Unfortunately, the Chechen forces had been long prepared for a strike against the city and the attack was a dismal failure. Reflecting failings of the 1979 Soviet invasion of Afghanistan, the loss of the initiative and surprise was compounded by a failure to conduct a proper assessment of the adversary. The Russians underestimated the will of the Chechens to defend their homeland, a failure accentuated by Nikolai Yegorov, the presidential special representative in Chechnya, who in 1994 questioned the inability of the Russian armed forces to subdue the republic: "do you mean to say that with our tanks we can't beat a load of shepherds?"[6] Instead of a few lightly armed farmers, the intervening Russian force was confronted with a large, well-armed enemy that was well established in defensive positions in an urban environment.

Challenges of Urban Combat

One of the key issues for the Russian forces was the lack of training in urban warfare: none of the units involved in the 1994 New Year's Eve storming of Grozny had received specialist training in urban warfare, which was rare in the Russian armed forces, in spite of their extensive experience of it during the Great Patriotic War, particularly in Stalingrad. According to General Mikhail Surkov, a former deputy chairman of the Duma Defense Committee who spoke about this issue in 1996, "street fighting tactics [we]re absent from the manuals of the Russian armed forces. Exercises simulating urban warfare are rarely carried out and our army has no experience of the real thing."[7] In contrast to the Russian troops, the Chechen fighters were lightly armed with machine guns, grenades, and grenade launchers and organized into small, highly mobile units. They were fighting on home ground and thus had a superior knowledge of the city layout. The federal forces were not even provided with suitable maps prior to the assault.[8] Valery A. Kiselyev and V. M.

Rybalko have used the Chechen conflicts as case studies to improve Russian practice in urban combat, noting the importance of underground infrastructure. They also emphasized the importance of ensuring troops have detailed, large-scale maps of cities and other urban areas, a deficiency that undermined the operational effectiveness of Russian troops during the first Chechen campaign.[9]

The Chechens used tactics that were simple but effective. They waited until the federal forces had reached the center of Grozny before separating the tanks from their infantry support and attacking them with grenade launchers and flame throwers.[10] Tanks and armored vehicles are not ideal in urban environments—their poor visibility and maneuverability mean that they are extremely vulnerable to attack, and they are reliant on support from infantry troops. In his testimony to the parliamentary commission investigating the causes of the war, a former deputy defense minister, Georgy Kondratyev, cast doubt over the deployment of tanks: "Why did they wage war with tanks in Groznyy? Tanks cannot do battle in cities."[11]

Despite adapting their tactics in the short term, the Russian forces failed to remember the crucial lessons of urban combat when they returned to Chechnya in 1999, and they took heavy casualties again when attempting to take control of Grozny. This emphasized their lack of preparation for urban combat. The Chechen conflicts witnessed the largest urban battles since the Great Patriotic War, and it is surprising that, despite a focus on historical experience, lessons from Stalingrad were overlooked. Russian forces sought to avoid direct fighting in the city and instead fought from a distance, using massive aerial and artillery bombardments to destroy it. The Russian Ground Forces have traditionally relied on artillery, the so-called "God of War."[12] According to Russian sources, precision weapons were used effectively at the beginning of the second Chechen campaign, although this did not prevent widespread destruction.[13] The psychological effect of artillery fires is perceived to be as great as their physical effects, and therefore they constitute part of Russian tactics to demoralize and undermine an adversary's will to resist. A number of analysts have stressed the importance that urban combat may play in twenty-first-century conflict, drawing on the experience in Chechnya to illustrate shortcomings in the Russian approach and lessons learned.[14] One of the key lessons that Russian military analysts have drawn from Syria is the central role of combat in populated, urban areas.

Coordinating Failure

Another key issue confronting the Russian intervention force was the challenge of coordinating between the numerous ministries and organizations who had provided troops. In addition to the Ministry of Defense, troops were also deployed by other power structures, including the Interior Ministry and Federal Security Service (FSB), who had little experience of operating together and were unable to coordinate or communicate properly with each other.[15] There was also poor integration between ground and air forces, leading to a number of friendly fire incidents.[16] In the three years between the wars, steps were taken to tackle the problem of command, control, and coordination. Training exercises were developed to prepare troops for combat within a joint force that had an integrated command structure, as well as to fight a low-intensity counterinsurgency operation against small "illegal armed formations" rather than a high-intensity conventional war against NATO. In 1998 an exercise was held in the North Caucasus and southern Russia involving around fifteen thousand troops from all of the power ministries to test cooperation and coordination among the varied Russian forces who were tasked with a range of missions from hostage rescues to urban combat and counterterrorism.[17]

Russia's experience in Chechnya emphasized the need for secure, modern command, control, and communications systems to ensure coordination between different force structures and ensure they are not vulnerable to interception by the adversary. Russian experts have been scathing about the absence of unified command and control structures during the interventions in Chechnya. Vorobyev asserts that one of the most serious and complex challenges faced by Russian forces in Chechnya was the absence of a unified command system that could organize interaction between groups of forces deployed by different ministries and security agencies.[18] He developed this further in a 2012 article, suggesting that lessons had still not been learned after decades of experience in Afghanistan and Chechnya, notably the absence of a single command structure for joint operations, which, in his opinion, is an "indispensable condition" for a successful operation, facilitating interconnection and interaction.[19] Aleksandr V. Khomutov pointed to communications problems experienced by units deployed in mountainous areas, where radio equipment did not function well, as well as the absence of coordination between ground forces and artillery and close-air support, a view echoed by a number of others, including Leonid S. Zolotov.[20]

Controlling Information Flows

One of the most important lessons that Russia learned during the 1994–1996 conflict was the crucial role of the media in modern warfare. At the beginning of the 1994 military operation, the Russian media retained a substantial amount of independence and were often openly critical of federal leadership's actions. During the first war, which was Russia's first televised war, the credibility of the Russian armed forces was consistently undermined by media reporting, which frequently contradicted the official position and demonstrated the brutality of the conflict to the Russian population. This played a major role in shifting public opinion against the invasion. Having gained an insight into the crucial role of the media, which was demonstrated during the NATO operation against Serbia in March 1999, the Russian leadership took resolute measures to ensure that the situation was not repeated. The attacks against apartment blocks across Russia in September 1999 (and the incursion into Dagestan by Chechen militant groups) provoked the Russian leadership to take decisive action against the alleged perpetrators and on October 1, 1999, then President Boris Yeltsin ordered troops to once again cross the borders into Chechnya to "eliminate terrorists." Putin, as prime minister, immediately took an uncompromising stance, vowing to chase terrorists "everywhere," including the "outhouse." A key element of this was the realization that a favorable result depended not only on success on the battlefield but also on success in the information war (*informatsionnaya voina*). The 1994–1996 war provided a clear lesson on the criticality of information flows in contemporary warfare.[21]

Consequently, during the second conflict, the flow of information was rigorously controlled by official sources,[22] while the government exerted considerable pressure on journalists and media organizations not to criticize or challenge policy.[23] A special information center was set up to hold regular press briefings to feed the official version of operations to the media. The Ministry for the Press, Television, and Radio Broadcasting issued a formal warning to all national radio and television networks against broadcasting any interviews with "Islamic rebel leaders" during the ongoing military conflict. The authorities argued that such interviews were in effect helping the militants to wage a "massive propaganda war" to incite ethnic and religious intolerance and promote a change in Russia's borders. The Kremlin also briefed Russian television heads on the "right" and "wrong" terms to be used in broadcasts about Chechnya. In an effort to highlight the international nature of the conflict, the term "Chechen terrorism" was to be replaced with "international terrorism," and the word "jamaat" (local Muslim community) was to be replaced with "terrorist organization or gang."[24] This was an attempt by the Russian leadership to situate its operation in Chechnya within a wider global context and counter international criticism of its actions.

A further reason for tighter Russian control of the flow of information was Putin's determination to shore up domestic support for a second Russian intervention. It had become clear during the first conflict that the Russian populace did not support punitive military action. Putin was determined to ensure that he had support from the Russian public for a second military operation to demonstrate resolve. Prior to launching the assault, the Russian authorities sought to gain unanimous support for a renewed military operation from both the political elites and the general public in a shrewd propaganda campaign. The Chechen people were vilified in this campaign, repeatedly referred to as "bandits" and "criminals" who threatened the stability of the Russian Federation. This demonization of an entire nation was conducted in order to win popular support for an invasion by depriving the Chechens and their cause of legitimacy.

At times, the Russian approach to information management was reminiscent of the Soviet era, whereas the approach of the Chechens demonstrated an appreciation for modern media and, in particular, the power of the internet. Militant groups turned to the internet to circumvent mainstream media, which they were denied access to. The technology of the internet played a key role in the information warfare waged by groups in the North Caucasus. The *Kavkazcentre* website was the mouthpiece for the more extremist groups, promoting global jihad and providing propaganda about attacks against "infidel" forces in conflict zones around the world. Shamil Basayev used the site in September 2004 to claim responsibility for the Beslan school massacre. The website was set up by the ideologist Movladi Udugov, a former Chechen "minister" for information and press, and proved very difficult to control—as soon as one country closed the site down, it moved to another. Aware of the importance of controlling information flows, the Russian authorities consistently sought to suppress Chechen rebel news outlets such as the *Kavkazcentre* website. Russian military theorists have been critical of the fact that Chechen militant groups were actively exploiting the internet to "promote its position, spread disinformation, attract funding and new mercenaries" through a wide range of outlets, whereas the Russian government position was promoted in a small number of official periodicals (including *Krasnaya Zvezda* and *Rossiiskaya Gazeta*) and websites.[25]

Working with Proxies

Russia provided covert assistance, both economic and military, to opposition groups and competing factions (including the Interim Council and its forces, led by Beslan Gantemirov) who were opposed to Dzhokhar Dudayev's rule, in an attempt to critically destabilize the regime without officially deploying Russian troops.[26] These efforts to undermine Chechnya through covert support for proxy forces ended with a failed intervention in November 1994, when proxy forces launched an abortive attack against the Chechen capital Grozny, supported by Russian military equipment, including tanks, armored vehicles and aircraft, and troops. Moscow refuted any involvement in the assault until several Russian newspapers published information from Russian officers claiming to have signed contracts with the FSK (predecessor of the FSB) "to take part in a secret military operation" in Chechnya.[27] This strategy of covertly seeking to exploit divisions between different factions to destabilize them was revived during the Second Chechen War and played a key role in Russia's approach to counterinsurgency, most notably the policy of "Chechenization." This was instigated in 2000 when Putin imposed direct rule over the republic and installed a pro-Moscow administration led by Akhmad Kadyrov, a mufti and former insurgent who had become increasingly concerned about the growing radicalization of many Chechen militant groups and had consequently changed sides.

GEORGIA

Unlike the counterinsurgency in Chechnya, Russia's five-day war with Georgia in August 2008 was a conventional military operation against another state actor, albeit one that incorporated elements of non-kinetic action, notably cyberattacks and information warfare. The war was also notable for the integration of different branches of the armed forces, including air, maritime, land, special forces, and airborne units under the command of a single headquarters. Russia maintains that its invasion on August 8, 2008, was merely a response to the Georgian attack on

Tskhinvali made to stop the alleged genocide of the Ossetian people by Georgian forces and to protect Russian citizens resident in South Ossetia, together with the Russian contingent of the Joint Peacekeeping Forces deployed there. The operation was referred to as an operation to "coerce Georgia to peace" (*operatsiya po prinuzhdeniyu Gruzii k miru*[28]), which would be realized by achieving the principal military objectives of securing control of Abkhazia and South Ossetia, and preventing Georgia from reinforcing by cutting off key transit routes.

To some extent, the war demonstrated the renewed ability of the Russian armed forces to fight conventional wars, following years of conflict in the North Caucasus. Russian mobility was certainly far better than it had been in previous conflicts: the deployed forces almost doubled in size within twenty-four hours even though Russia could not begin an immediate airlift because of a failure to suppress the Georgian air defenses and the bottleneck that formed in the Roki Tunnel (through the Caucasus mountains, a direct link between Russia and South Ossetia), which forced a slow pace. Within a few days, as many as ten thousand Russian troops had advanced into South Ossetia and up to nine thousand in Abkhazia.[29] Russia's military advantage was reinforced through the use of proxy forces, an enduring feature of all Russia's post-Soviet interventions. The International Commission that investigated the August 2008 conflict described it as a combination of an interstate and intrastate conflict, involving military engagements between regular armed forces, as well as armed actions by militias and irregular armed groups.[30] South Ossetian and Abkhazian militias (*opolchentsy*) supported Russian troops, and there were also reports of Chechens and Cossacks taking part in the conflict.

There is evidence to suggest that the Russian operation was long planned, with some suggesting that such a large-scale invasion involving a joint force would have required considerable preparations. It has been suggested by some Russian analysts that Moscow was aware of Georgian plans for an attack against South Ossetia; it just did not know the precise date.[31] Consequently, it is possible that the Russian military command had made preparations to be able to intervene in South Ossetia as soon as possible after the beginning of any Georgian offensive. In spite of this, Moscow again did not hold the initiative during the opening phase of the conflict, which began when Tbilisi attempted to restore central Georgian control over South Ossetia on August 8, 2008, triggering the Russian military invasion. Nevertheless, Russia did have the element of surprise (discussed later) and quickly regained the initiative.

Taking the Georgians by Surprise
Despite the fact that Moscow did not hold the initiative during the IPW, it still used the element of surprise to full effect: the Georgians were totally unprepared for a large-scale Russian military intervention and had misinterpreted the Kremlin's intentions. The Georgian armed forces were prepared for a mobile, offensive war against separatist forces in either South Ossetia or Abkhazia, not for simultaneous large-scale combat against tens of thousands of well-armed and well-supported Russian troops invading from both South Ossetia and Abkhazia—that is, two fronts at the same time. In his testimony to the parliamentary commission investigating the war, Alexander Lomaia, the secretary of Georgia's National Security Council during the war, claimed that Russia used around one-third of its combat-ready ground forces in the operation against Georgia and that "neither we nor any foreign intelligence service had any information about Russia's expected full-scale invasion and occupation of a large part of our territory—it was a shock and a surprise."[32]

The Georgian government and military were stunned and confused by the speed of the Russian invasion, rendering them unable to stage any meaningful resistance. Russian actions appear to reflect Aleksandr Suvorov's principle of "to astonish is to vanquish" (*udivit—znachit pobedit*).[33] A key factor in the speed of the Russian military victory was the opening of a second front in Abkhazia, along with the severing of the main transportation routes in Georgia. Alexander Lomaia testified that the probability of such a situation where a significant proportion of Georgian territory, including key routes, was controlled by Russia, was at the bottom of the list of possible scenarios that had been considered.[34] Georgia's east-west "Caucasian Corridor," the central valleys that run between the High Caucasus to the north and the Lesser Caucasus in the south, is narrow, squeezed between the two mountain ranges. It is through this narrow corridor that the principal transport and communications links run, including pipeline infrastructure, highways, and rail routes, giving the Russians easy access to a range of Georgia's critical national infrastructure. The Russians used strategic surprise, combined with speed of action, to great effect, enabling them to rapidly regain the initiative and achieve their objectives.

Ongoing Communications and Coordination Challenges

In spite of overwhelming numerical superiority and the rapid expulsion of the Georgian armed forces from South Ossetia, Russia's military performance in the 2008 war highlighted some significant weaknesses in capability, notably ongoing problems with command, control, and communications, poor air/land integration, and obsolete, unreliable equipment. As Deputy Defense Minister General Nikolai Pankov stated in 2009:

> The operation to force Georgia to peace has clearly shown that...
> the army is ready to solve its tasks in any situation. We gained
> some experience in warfare, saw "bottlenecks" in the training of
> our military personnel. Of course, we have learned from this.[35]

In spite of the lessons from Chechnya, one of the key problems identified during the Georgian conflict was obsolete communications systems, making it difficult to rapidly transmit intelligence and effectively manage units on the ground. The most obvious example of this was the fact that Russian commanders had to resort to using the mobile and satellite phones of journalists in order to be able to communicate with individual units that were farther forward.[36] Furthermore, interaction between Russian Ground Forces and the Air Force was limited by communications issues and a lack of coordination. Viktor Kutishchev notes that the formation of a truly integrated group of forces, facilitated by good communications and coordination, is an "essential condition" for conducting operations in accordance with the principles of network-centric warfare.[37] He went on to point out the discrepancy between the theory and reality of contemporary conflict, arguing that the Russian armed forces were still focused on outdated views of traditional, large-scale ground operations rather than "modern concepts that involve the widespread use of precision weapons." The Russian armed forces deployed to Georgia did not have precision munitions, despite their use being a central focus of Russian military theorists since the 1990s. They also did not appear to use UAVs. While Russian military thought had evolved through observation of Western interventions, the procurement process and doctrinal change lagged behind. The conflict exposed enduring problems outlined above and led leaders to the conclusion that Russia was poorly prepared to fight a modern conflict, even against a weaker opponent. This prompted a further round of ambitious reform by former Minister of Defense Anatoly Serdyukhov, who sought to get rid of the outdated concept of mass mobilization, based around divisions, and replace it with high-readiness brigades comprising predominantly professional, or "contract," soldiers. His military reform program led to further cuts in both personnel and equipment, with the aim of creating a more mobile military with an expeditionary focus.[38]

Critical Role of the Information Campaign

Information operations and warfare played a leading role in the conflict, although there was a noticeable shift in the way that Russia sought to use the information sphere in comparison to the Chechen conflicts discussed previously. In Chechnya, particularly during the first conflict, information was perceived to have hindered kinetic activity, undermining the Russian military position. During the 2008 conflict in Georgia, non-kinetic actions were synchronized with military action at the tactical and strategic level, intended to both complement and amplify kinetic operations on the ground. There were two key elements to Russian information operations:

cyberattacks against Georgian websites, particularly those linked to the government, and attempts to control the narrative. There were reports of distributed denial-of-service (DDoS) attacks against Georgian sites in July 2008, several weeks before the kinetic invasion, with one report highlighting a stream of data directed at government sites containing the message "win+love+in+Rusia."[39] Servers used by government and media sites, including those used by the National Bank of Georgia and the president's website, were targeted, in an apparent attempt to undermine the government and prevent it communicating clearly, forcing it to post updates on a blogging site.[40] In addition to DDoS attacks, some Georgian sites were also defaced, most notably with pictures that compared Georgian president Mikhel Saakashvili with Adolf Hitler, in an attempt to divide the Georgian population and undermine their will to resist.[41] The incorporation of cyberattacks into a broader kinetic war signified a step change in the Russian approach, a manifestation of the significance accorded to information warfare, of which cyber is but one element, as well as the use of all available means in pursuit of strategic objectives.

The second key element of Russian information operations was the focus on ensuring that the Russian narrative regarding the conflict was able to compete with both Georgian and Western narratives. Building on lessons from Chechnya, leaders focused information warfare not just on the domestic constituency but also on international opinion and foreign audiences. Information warfare is perceived to be a powerful means of indirect, nonmilitary influence over open societies, partly because it is very difficult to attribute. Some Russian observers discern a growing confrontation in the global information space driven by the US and its allies, who are seen to be "manipulating the public consciousness" of their own domestic populations and those of their adversaries.[42] Thus, the control of narratives at home and abroad is considered to be of critical importance for Russian security. Russia learned a tough lesson in Georgia in 2008, losing the information war—despite wielding overwhelming military power—to President Saakashvili, a US-educated lawyer, who made frequent international television appearances, appealing in English for Western support in the face of Russian aggression. Saakashvili also made reference to Russian brutality in Chechnya, saying Russia must not be allowed to turn Georgia "into another Grozny."[43]

Notwithstanding some apparent success in the information arena, concern was expressed by some analysts about Russia's inadequacies in the strategic communications arena, especially vis-à-vis the US, which was perceived to have deployed its full range of strategic communications tools in support of Georgia in order to undermine Moscow. In addition to assisting senior Georgian officials in cooperating with international media and giving interviews in English, Russian analysts accused the US of using the UN Security Council as a forum to promulgate pro-Georgian opinions, encouraging the shutting down of Russian TV channels in Georgia and spreading misinformation regarding who had initiated the conflict, including information from the conflict zone, notably Russian actions toward Georgian civilians, the destruction of cities, and Russian losses.[44] As a result of the lessons learnt in Georgia, the Russian leadership invested in instruments of information warfare (to be discussed in chapter 5), including promoting the Russian language, developing a heavy international presence for Russian media, and honing its strategic communications skills to ensure a unified narrative across government. In its Foreign Policy Concept of 2013, Russia declared that it must "create instruments for influencing how it is perceived in the world," "develop its own effective means of information influence on public opinion abroad," and "counteract information threats to its sovereignty and security."[45] In line with these goals, the Russian government has invested in media resources that can convey its point of view to other countries, such as the TV news channel *RT* and the news agency *Sputnik*. Information confrontation went on to play a central role in Russia's intervention in Ukraine.

UKRAINE

Russia's swift annexation of Crimea in March 2014 and support for pro-Russian separatists in eastern Ukraine elicited extensive international focus on Russia's supposedly new approach to warfare, so-called "hybrid warfare." This became the defining label for Russian intervention in Ukraine prior to its full-scale invasion in 2022, and a concept seen at the time by some observers as a potential model for future conflicts involving Russia. Although the impulses for Russian intervention in Ukraine were similar to its 2008 war with Georgia—a desire to prevent the state developing closer ties with Western organizations and moving away from Moscow's sphere of influence—Russia's military involvement has been very different, driven by a different strategic logic and circumstances. Intervention has been portrayed by Moscow as a defensive step to protect itself (and Ukrainians) from the instability caused by a Western-sponsored "coup": in a 2015 speech, Gerasimov accused the West of "curating" events in Ukraine, warning that the uncertainty surrounding events could lead to a military threat to Russia.[46] One of the principal objectives in Ukraine was coercing Kyiv while maintaining an element of deniability and concealment. Rather than launching a large-scale military invasion supported by air power, as in Georgia 2008, Russia relied upon long-range artillery and rocket fires, electronic warfare, and cyber capabilities, together with information operations and proxy forces, in order to conceal its participation and facilitate deniability. Information has also played a central role in the Russian approach, revealing significant advance from 2008. It is important to differentiate between Russia's Crimean operation and its support for separatists in eastern Ukraine. The Crimea operation was completed swiftly, whereas Russian involvement in eastern Ukraine continued until its invasion in February 2022. Furthermore, although the Russian government continued to deny its presence in Donbas, Putin admitted Russia's involvement in Crimea in the documentary *Krym: put na rodinu*;[47] this ambiguity makes it harder to assess the lessons learned in Ukraine, although it is possible to trace change and continuity with previous operational experience. There is an absence of publicly available material that evaluates Russia's experience in Ukraine; unlike operations in Chechnya, Georgia, or Syria, there have been no public roundtables to discuss lessons learned, and articles in the military theoretical press tended to focus on Western assistance to and involvement in Ukraine.

Taking Crimea by Surprise

Surprise and seizure of the initiative played a key role in Russia's success in Crimea. The speed of the Russian deployment in the spring of 2014 took the Ukrainian government by surprise, undermining its ability to make decisions and offer any resistance. The rapidity of the intervention also took the international community by surprise, forestalling any unified response: while the world's attention was focused on the Winter Olympic Games in Sochi, Moscow acted quickly to take decisive control of the Crimean peninsula, with offensive action that denied Ukraine the initiative and sowed confusion. Furthermore, the Russian approach was focused on ambiguity in order to confuse and forestall any international response. Russian troops were given the label the "polite people": a euphemism for the heavily armed and unidentified soldiers, "the little green men," who took over Crimea. Putin consistently denied that the "little green men" had anything to do with Russia. Although the Russian president admitted in 2015 in *Krym: put na rodinu*[48] on the first anniversary of the annexation of Crimea that he had ordered the operation, in the spring of 2014, the ambiguity surrounding the "little green men" served a clear purpose: to confuse the adversary and degrade their capacity to calculate and act in a coherent fashion, thus undermining their will to resist. Another key aspect of the Crimea operation was the minimal use of force, which demonstrated a noticeable shift in Russian thinking. Prior to 2014, Russian operations tended to rely heavily on the large-scale use of force: operations in Chechnya and Georgia were characterized by a reliance on the use of overwhelming military firepower, particularly Chris Bellamy's so-called "God of War," artillery. In contrast, the operation in Crimea was conducted with very limited violence and bloodshed, largely because there was little resistance from Ukrainian forces as a result of government disarray in Kyiv, as well as the presence of Russian troops in Crimea prior to 2014. Furthermore, Russian air power also played a very limited role in Ukraine, partly because of the Russian desire for deniability. Nevertheless, the threat of conventional forces was used to pressure the government in Kyiv and shape its decision-making. It also used military exercises to mass forces on the border between Russia and Ukraine, either as a thinly veiled demonstration of military might or as cover for a concentration of forces prior to an invasion.

Relying on Proxy Forces in Eastern Ukraine

Russia's intervention in eastern Ukraine (2014 to early 2022) was based on a very different approach than that used in Crimea, in part because the element of surprise (and therefore the ability to seize the initiative) was no longer available: the Crimean operation had relied very heavily on surprise, but, following its success, further Russian intervention in other parts of Ukraine (and elsewhere) was expected. Paralleling its approach in Chechnya prior to the December 1994 invasion, Moscow relied heavily on the use of irregular forces and proxy groups to destabilize the situation in the Donetsk and Luhansk *oblasts*, resulting in the emergence of an externally supported separatist insurgency. The Kremlin attempted to undermine Ukraine from within through the use of political warfare and subversion but underestimated Ukraine's will to resist, as well as its own ability to develop an effective network of proxy forces in eastern Ukraine. The establishment of the National Defense Management Center (discussed in more detail in chapter 4), which was put into operation in December 2014, improved the ability of Russia's defense and security structures to manage operations and disparate groups, facilitating better sharing of information and situational awareness. Nevertheless, Moscow continued to deny its involvement in the situation in eastern Ukraine. It also continued to use conventional forces to distract and deter, with large-scale military exercises taking place close to Russia's border with Ukraine.

Information Confrontation

Russia built on its operational experience in Georgia of combining kinetic and non-kinetic actions, using cyberattacks and information operations. As will be discussed in chapter 5, the Russian definition of information warfare ("information confrontation" in Russian) categorizes the technical and psychological aspects of the approach as discrete elements. Information confrontation has been central to the Russian approach in Ukraine to confuse and forestall any response. The fiber-optic cables between the Crimean peninsula and mainland Ukraine were severed in late February 2014, disrupting communications, and advanced electronic warfare capabilities were used to degrade Ukrainian command and control while masking Russian operations.[49] The Russian leadership also sought to justify its annexation of Crimea and support for separatists, honing its expertise in strategic communications in recognition of the paramount importance of shaping and controlling the information space, emphasized in a speech by Collective Security Treaty Organization chairman Nikolai Bordyuzha at the Moscow International Security Conference in May 2014.[50] In the run-up to its annexation of Crimea, Roskomnadzor, the federal regulator, blocked access to independent, pro-opposition news sites such as *Dozhd*, as well as the sites of Kremlin critics. Russia took a far harsher line following its invasion of Ukraine in 2022, seeking to ensure that it had almost total control of information flows within Russia. To ensure that the state narrative prevailed, it strengthened its control over domestic media, attempting to prevent anti-war protests threatening internal stability.[51]

SYRIA

Russia's activities in Syria were indicative of a substantive change in its approach, reflecting the lessons it had learned over the preceding decade or so, both from observing others and its own operational experience. Russian theorists and experts continue to analyze and assess the conflict for indications of the evolution of conflict, viewing it as a potential prototype of future war. Anatoly D. Tsyganok deems Russia's Syrian campaign to constitute a "new type" of war, characteristic of the modern era, combining a wide range of instruments, including political, psychological, and informational. The military aspects remained limited, largely the result of the wish to avoid large-scale Russian losses.[52] Gerasimov believes that the combat experience gained in Syria has accelerated the development of new ways of using the Armed Forces, as well as new forms and methods of operation.[53] Shoigu, the defense minister, has echoed this, asserting that Russian troops had to learn to fight in a new way in Syria, "and [that] we have learned this."[54] In contrast to the campaign in Ukraine, Russia's operations in Syria have been portrayed as an expeditionary, non-contact, Western-style war, using precision guided weapons, and conducted primarily by the Aerospace Forces, with only limited numbers of ground forces deployed, a significant shift from previous operations. There has been increased use of proxy forces, including local militias and PMSCs, and the campaign has provided a vital testing ground for military personnel and weapons. Syria has also seen the Russian forces integrating technological enablers such as autonomous and robotic systems into their existing systems and approaches in order to undermine an adversary's command and control and communications. Referring to operations in Syria, Gerasimov noted in 2018 that some experts have identified it as "new-generation war," stating:

> Changing the character of conflict is an ongoing process..., all recent military conflicts have been significantly different from each other. The very content of military operations is also changing. Their spatial scope is increasing, along with tension and dynamism. The time parameters for preparing and conducting operations are reduced.[55]

Although Gerasimov recognizes the invaluable combat experience that the Russian armed forces have gained in Syria, he has been critical of military science for failing to capitalize on this experience and analyze the lessons that can be drawn. In an article in *VPK* in 2017, he called for real results rather than "puffed-up" reports:

Military science has always been distinguished by the ability to see and reveal problems at the stage of their appearance, and the ability to work them out quickly. Unfortunately, this quality has been lost. Now, when solving practical problems, military command and control bodies do not always have the opportunity to rely on the results of their preliminary scientific study. Therefore, one of the tasks of these bodies for the management of military science is to ensure the relevance of research, to create the necessary conditions for this....It is especially important that the results of research are quickly translated into the practice of the Armed Forces.[56]

He encouraged theorists and experts to focus on "promising new areas," notably combat in space and the information arena, advanced weapons and control systems, and the development of forms of strategic actions. His speech represents another call to arms for Russian military scientists, exhorting them to systematically assess the evolution of conflict and the implications for Russia.

In the opinion of Russian experts, the military's experience in Syria demonstrated that there has been a transition from military actions that are sequential and concentrated to ones that are continuous and distributed, conducted simultaneously "in all spheres of confrontation," requiring increased mobility. An adversary will be defeated through the integration of all strike (long-range, operational) and fire (tactical) weapons into a single system; the role of electronic warfare, information confrontation, robotics, and UAVs is also growing, increasing an actor's ability to influence the enemy at distance: "in a complex, rapidly changing environment, the ability to effectively control troops and forces is of particular importance."[57] General Colonel Aleksandr V. Dvornikov, commander of the Southern Military District and a former commander of operations in Syria, has offered several insights into the changing nature of military art, emphasizing the use of "integrated" formations and the growing importance of information warfare as the most important issues. He also noted the use of "lower-level forms" (operations, battles, strikes) and the erasing of boundaries between strategic, operational, and tactical tasks: "strategic (operational) goals were achieved by the actions of military formations of the tactical level."[58] In his view, integrated units had a number of features, including autonomy and mobility; the complex use of military force that involved the armed forces of several states, alongside militia groups; the use of long-range strikes to "reduce the economic potential" of the adversary; and the use of guerrilla tactics as well as classical forms of combat.[59] He also noted that humanitarian operations represented a new form of operation for the Russian armed forces in Syria, as well as the key roles played by Special Operations Forces and Russian military advisors working with the Syrian armed forces. It is notable that Dvornikov considers humanitarian operations and post-conflict resettlement to be a new form of deployment for the Russian armed forces, not activities that they had engaged in during the years spent in Chechnya and the wider North Caucasus.

Precision Strike

One of the most notable aspects of the Russian operation in Syria has been its use of long-range high-precision weapons: air and sea-based missiles, as well as long-range bombers, have been used to destroy targets from distance. The use of Kalibr missile strikes from corvettes in the Caspian Sea to hit targets over 1,500 kilometers away in Syria for the first time in October 2015 demonstrated the evolution of Russian thought and new capabilities in precision strike. In a 2017 speech, Gerasimov underscored the extensive use of precision weapons in Syria to deter and defeat, noting that precision strike had become an integral part of all military operations.[60] He also emphasized the use of reconnaissance-strike and reconnaissance-fire tactics, noting that the widespread use of reconnaissance and strike capabilities, based on intelligence, control, and communications, has led to the implementation of a "one target, one bomb" principle.[61] Shoigu has called for the development of ground-based alternatives to the sea-based Kalibr, which he says "proved itself" in Syria. The defense minister reflected upon some of the lessons from Syria during a meeting with Russian military leaders in 2019, emphasizing the necessity of detailed reconnaissance and cartographic information in order to be able to make effective use of high-precision weapons systems.[62] The role of satellites and space-based assets has become increasingly important for Russia and Russian military thought, reflecting the evolution of Soviet military thought.

Testing Weapons and Troops

Syria provided an important testing ground for a number of new Russian weapon systems and military equipment, including sea and air-launched cruise missiles, aircraft and helicopters, and reconnaissance and targeting systems. The challenges of combat and the difficult climatic conditions in Syria provided an environment for testing that was deemed far superior to domestic testing grounds. Furthermore, defense industry representatives were stationed at Russia's Hmeimim base with repair squads to carry out necessary modifications and repairs on the ground.[63] Gerasimov maintains that the experience in Syria proves that Russian weapons and equipment are "among the best in the world," easy to operate and reliable, suggesting that, as a direct result of the Syrian example, many countries were now looking to purchase Russian arms.[64] This optimism may have been eroded by the Russian military's performance in Ukraine following its invasion in February 2022, which emphasized the flaws and vulnerabilities of certain Russian weapons systems and equipment.

The key role of communications and reconnaissance has also been accentuated during the Syrian operation, according to a number of Russian analyses. One of the principal lessons drawn from Russia's experience in Syria has been the criticality of command and control, ensuring "superiority of management" of an integrated force deployed on a multi-domain operation. According to General Aleksandr Lapin, commander of the Central Military District, "competent and continuous management is the main guarantee" of a successful operation.[65] Colonel Viktor Tagirov, the chief of communications and deputy chief of staff of the combined-arms army for communications of the Western Military District, provided his assessment of the improvements in Russian signals capabilities, noting that one of the principal elements of the experience in Syria was the ability of crews to operate independently, separate from main bases and units. In his view, modern Russian communications technology has made a "decisive leap forward," boosting command and control and, consequently, significantly increasing combat capability. He pointed to the fact that the technology was tested by extreme climatic conditions during combat in Syria (as well as tests in the Arctic) and had proven itself to be very reliable.[66] Problems did arise with the creation of a satellite communications network: Lieutenant General Khalil Arslanov from the General Communications Directorate noted particular difficulties caused by low bandwidth and "dead zones." Once these issues were resolved, the ground grouping of satellite communication stations increased threefold, the capacity of repeaters increased, and the antenna systems of military and dual-use communications satellites were reoriented. The Syrian operation was notable because it was the first time that the use of foreign communication networks by wireless broadband equipment was tested and implemented as a means of ensuring classified communications.[67]

The Russian intervention has also provided invaluable combat experience for the country's military commanders, who have been rotated in and out of the campaign continually for the past several years, along with teaching staff from the majority of the professional military educational institutions. For some military leaders, one of the most important impacts of the ongoing operation for the Russian Armed Forces is the exchange of operational experience gained in Syria and the implementation of specific training that incorporates the lessons from this experience.[68] The commander of the Central Military District, Lieutenant General Aleksandr Lapin, stated that the Vostok-2018 exercises included "new forms and methods of combat" based on Russia's operational experience in Syria, a statement he repeated in 2020, when exercises included multi-domain operations, necessitating an integrated force that included electronic warfare, reconnaissance, air defense, and long-range precision-strike.[69] General Dvornikov underscored the shift in emphasis of the annual Kavkaz exercises, which in 2016 focused on localized armed conflict on

Russia's periphery (in this case the Black Sea region), but in 2020 were based on a Syrian-type conflict—that is, an international armed conflict involving foreign troops—in order to integrate lessons from the "combat experience of modern armed conflicts."[70] Military exercises are important indicators of a state's intentions for its armed forces—what it wants them to be able to do. The commander of Russia's Western Military District, General Aleksandr Zhuravlev, has stressed the significance of the experience acquired by the armed forces in Syria, saying it could not be overestimated, and set out how it was being incorporated into training exercises. In 2019, training focused on urban combat, including "tunnel warfare," along with the use of new "non-standard forms and methods of action" involving electronic warfare, reconnaissance-fire and reconnaissance-strike systems, and integrated operations involving ground forces working in cooperation with aviation and UAVs.[71] He also emphasized the need for creative thinking amongst officers:

> Modern combat, as confirmed by the Syrian experience, requires commanders of all levels to display military ingenuity, to seek opportunities to mislead the enemy, to force him to act in a manner favourable to us. Consequently, greater attention is paid to the advanced readiness of control systems...as well as...covert deployments.[72]

A critical feature of the Syrian conflict is the fact that the majority of fighting has taken place in urban, populated areas. Militant groups such as ISIS have been using underground tunnels, reflecting Russia's experience in Chechnya. Furthermore, Colonel General Igor Korobov, head of the General Directorate, noted that terrorists use so-called swarm tactics, the essence of which is to organize chaotic shelling and surprise attacks on units of the Syrian army by small but well-coordinated groups. Continuous attack exhausts, demoralizes, and deprives large military units of their initiative. These issues were all discussed in detail at a series of roundtable discussions organised in 2017 by the Military Academy of the General Staff to analyze and assess the Syrian experience.[73] Russia's Syrian operation has also been notable for the prevalent use of proxy forces, including private military companies, which will be discussed in more detail in chapter 6. Proxy forces, including local and Iranian-backed militias such as Hezbollah, have been responsible for ground operations while Russia provided close-air support, but there have been significant challenges in ensuring effective cooperation between the Russian forces and proxy groups.[74] Syria has also constituted a key testing ground for the Russian use of PMSCs during military operations, providing an active military force that cannot be linked to the Russian Armed Forces and thus providing the Kremlin with deniability, discussed in more detail in chapter 6.

Strategies of Limited Action and Active Defense

Gerasimov built on previous statements about Syria in his annual address to the Academy of Military Sciences in 2019. He outlined new strategies that had developed directly from the Russian intervention in Syria, notably the strategy of limited actions and the strategy of active defense (see chapter 5 for more detail on this). He noted that the Syrian experience had an "important role for the development of strategy...carrying out tasks to defend and advance national interests outside the borders of Russian territory within the framework of the 'strategy of limited actions.'"[75] In order to implement this strategy, Gerasimov said that it was necessary to gain information superiority, ensure effective command and control, and use covert deployments as well as a self-sufficient, highly mobile group of forces. Lessons from, and practical experience gained during, the Syrian operation are being incorporated into the curriculum at professional military education establishments such as the Military Academy of the General Staff, reflecting a belief that "science without practice is dead."[76]

A significant element of the strategy of limited actions is the conduct and coordination of military and nonmilitary actions by both Russian forces and the armed forces of other states, as well as militia groups, necessitating the "creation and development of a unified system of integrated forces and means of reconnaissance, destruction and command and control of troops and weapons on the basis of modern information and telecommunication technologies...In the future, military science needs to develop and justify a systems complex defeat of the enemy."[77] Gerasimov called for the widespread use of robotic systems during operations, particularly UAVs, to improve military efficiency, along with the creation of electronic warfare systems to counter an adversary's use of UAVs and high-precision weapons. Gerasimov's speech accentuates Russian thinking on the changing character of conflict, which envisions digital technologies, AI, unmanned systems, and electronic warfare at the heart of contemporary conflict.

Twenty years after its early failures in Chechnya, Russia's ambiguous use of force in Ukraine was seen by many to represent significant change in the way that Moscow wields its military power. In contrast, its intervention in Syria revealed a qualitative change in the approach of the Russian armed forces, as well as the weapons and military equipment deployed, reflecting lessons learned over two decades. Nevertheless, despite a belief that Russian operations in Syria constituted a new approach, there were a number of echoes of Russia's experience in Chechnya, particularly a reliance on intense bombing and indiscriminate airstrikes against residential areas and civilian infrastructure such as schools, hospitals, and markets. Despite the widespread Western conviction that Russian activities in Ukraine were evidence of a significant change in approach, it has actually demonstrated considerable continuity. It is clear that Russia's interventions in Ukraine (pre-2022) and Syria demonstrated the results of processes that have been ongoing in the Russian military and in Russian strategic thought over the past few decades, driven partly by Russian perceptions and understanding of the military activity. There has been significant continuity in some areas, notably a focus on the need to seize and maintain the strategic initiative in order to dominate the tempo of a conflict, prompting an emphasis on surprise and deception, as well as the ongoing importance of the initial period of war and the use of all available means, including a consistent focus on conventional forces. There has been a focus on the use of nonmilitary means as a way to exploit adversary's vulnerabilities, as well as special forces. In the Russian view, wars are to be dominated by information and psychological warfare in order to achieve superiority in troops and weapons control, undermining an opponent's will to resist. The main objective is to reduce the need to deploy hard military power to the minimum necessary. Although recent operations have demonstrated significant continuity in the Russian approach, there have been some important changes, including greater integration, a whole-of-government approach reflected in the establishment of the NTsUO. The following sections will examine some of these areas in more depth, analyzing military thinking on the impact of technological advances, the growing role of nonmilitary means of achieving strategic objectives, and the evolution of the Russian approach to proxy forces, particularly the use of PMSCs.

NOTES

1. The Soviets counted on the surprise shock of the initial invasion and short-term military occupation to undermine their adversary but had underestimated the resolve of the Afghan population and its will to resist. For an in-depth analysis of the Soviet experience in Afghanistan see Olivier Roy, "The Lessons of the Soviet/Afghan War," *Adelphi Papers* 259 (Summer 1991), London, International Institute for Strategic Studies. Aleksandr Liakhovskii, *Tragedia i Doblest Afghana* (Moscow: GPI Iskona, 1995); Carl van Dyke, "Kabul To Grozny: A Critique of Soviet (Russian) Counter-Insurgency Doctrine" *The Journal of Slavic Military Studies* 9, no. 4 (December 1996): 689–705.
2. Arbatov, "The Transformation of Russian Military Doctrine: Lessons Learned from Kosovo and Chechnya," v.
3. Eugene Miakinkov, "The Agency of Force in Asymmetrical Warfare and Counterinsurgency: The Case of Chechnya," *Journal of Strategic Studies* 34, no. 5 (2011), 667.
4. The northern column advanced from Mozdok in North Ossetia (where operational headquarters were based), the western column crossed Ingushetia from Vladikavkaz in the west, and the eastern column started in Kizlyar, Dagestan to the east. Estimates vary as to the size of the combined force. Pavel Grachev, the Russian defense minister, maintained that for the first stage of the operation "a group of forces with an overall strength of 23,800 officers and men was created." The group included 19,000 troops of the Armed Forces, 4,700 Interior Ministry troops, 80 tanks, 208 ICVs and APCs [armored troop carriers], and 182 artillery pieces and mortars. For the Russian order of battle and details of the composition of each battle group see Charles Blandy, David Isby, David Markov, and Steven J Zaloga, "Jane's Intelligence Review Special Report No 11," in *The Chechen Conflict: A microcosm of the Russian Army's past, present and future* (1996), 12–13.
5. Colonel-General Boris Gromov, Valery Mironov, and Georgy Kondratyev were all rumored to have been dismissed because of their criticism of the Chechen campaign. See *Izvestiya*, December 10, 1994, 4, quoted in Tracey German, Russia's Chechen War (London: Routledge-Curzon, 2003), 132. Other senior commanders resigned their government posts in protest at the invasion. Major General Aleksandr

Tsalko and Colonel Vladimir Smirnov left their posts at federal power bodies in mid-December. See the Summary of World Broadcasts database SWB SU/2188 B/2 [2] 29.12.94 - *Kuranty*, Moscow, 2.12.94.

6. Cited in C. W. Blandy, *Chechnya: Two Federal Interventions. An Interim Comparison and Assessment*, report no. P29, Conflict Studies Research Center, January 2000, 13.

7. Quoted in Tracey German, *Russia's Chechen War* (London: Routledge-Curzon, 2003), 133.

8. Lieutenant-Colonel A. Frolov, who was based in the northern area of operations at the beginning of the conflict, recounts how commanders were only issued with maps on a scale of 1:100,000, rather than the 1:25,000 maps that are vital when attacking a city. Aleksandr Frolov, "Soldaty v avangarde i komandiry v Mozdoke," *Izvestiya* (January 11, 1995): 4.

9. Valery A. Kiselyev and V. M. Rybalko, "Ob ispol'zovanii podzemnykh kommunikatsii i sooruzhenii pri vedenii boyav gorode," *Voennaya mysl'* no. 1, (2002): 34–38.

10. For a detailed eye-witness account of the storming of Grozny, see Vadim Belykh et al, "Voennyie i separatisty," *Izvestiya* (January 6, 1995), 2. The rebel fighters would block the streets with burning cars in order to halt the advancing columns of Russian troops. Snipers then fired grenades from nearby houses, aware that the poorly armored APCs had additional fuel tanks by the rear doors. The crew either abandoned the burning vehicle to be shot at by snipers or remained inside.

11. Originally trained in a tank regiment, he was highly critical of the fact that the Russian tanks were supported and covered by infantry rather than vice versa. G. Anishchenko, A. Vasilevskaya, O. Kugusheva, and O. Mramornov, *Kommissiya Govorukhina* (Moscow: Laventa, 1995), 98.

12. Chris Bellamy, *Red God of War: Soviet Artillery and Rocket Forces* (London: Brassey's, 1986).

13. Slipchenko in Gareev and Slipchenko, *Future War*, 41.

14. See for example A. F. Bulatov, "Sposobyi ovladeniya gorodami i usloviya ikh primeneniya," *Voennaya mysl'* no. 2 (2002): 23–28.

15. Olga Oliker, *Russia's Chechen Wars 1994–2000: lessons from urban combat* (Santa Monica: RAND, 2001), 14.

16. One estimate puts friendly fire casualties as high as 60 percent of total Russian casualties in Chechnya. N. N. Novichkov, V. Snegovskii, A. G. Sokolov, and B. Shcvarev, *Rossiisskiye vooruzhennyye sily v chechenskom konflikte: analiz, itogi vyvidy* (Paris: Kholveg, 1995), 70, citing an

unnamed counterintelligence officer quoted in an *Izvestia* article from February 15, 1995.

17. Oliker, *Russia's Chechen Wars 1994–2000*, 37.

18. Ivan Vorobyev, "Vzaimodeistvie silovykh struktur v vooruzhennom konflikte," *Voennaya mysl'* no. 6 (November 1999).

19. Ivan Vorobyev and Valery A. Kiselyev, "Politseisko-voiskovaya operat-siya" *Armeiskii Sbornik* no. 10 (October 2012), 59–61.

20. Aleksandr V. Khomutov, "Boi vedyet batalyonnaya takticheskaya gruppa," *Armeiskii Sbornik* no.6 (June 2009): 23–26; Leonid S. Zolotov, "Kontrterroristicheskaya operatsiya v Dagestane i Chechnye: osnovnyie itogi i vyivody," *Voennaya mysl'* no. 3, (May 2000).

21. See for example, D. P. Prudnikov, "K voprosu ob informatsionnoi sostavlyayushchei voenno-upravlencheskoi deyatelnosti v sovremen-nykh usloviyakh," *Voennaya mysl'* no. 4 (April 2008): 23–28. Prudnikov analyzed Russian television coverage of the 1994–1996 conflict.

22. Accreditation was introduced, journalists were accompanied to the scenes of events at all times by military personnel, and contacts with separatist forces were cut off. For further details, see Olessia Koltsova, *News, Media and Power in Russia* (London: RoutledgeCurzon, 2006), 205–225.

23. Television stations that take an independent line of the Kremlin have been subject to strong legal pressures such as criminal or tax investiga-tions that have made it difficult for them to operate as independent sta-tions. Ensuring press freedom is even more difficult in Russia's regions where media outlets are heavily dependent on authorities for financial subsidies and are particularly vulnerable to harassment and intimida-tion.

24. Igor Torbakov, "War on terrorism in the Caucasus: Russia breeds jihadists" *Chechnya Weekly* 6, no. 42 (November 10, 2005), The Jamestown Foundation, https://www.jamestown.org.

25. Yu. O. Yashchenko, "Internet i informatsionnoe protivoborstvo," *Voen-naya mysl'* no. 3 (2003): 76–77. Also Oleg N. Kalinovsky, "Informatsion-naya voina – eto voina?," *Voennaya mysl'* no. 1 (2001): 57–59.

26. For further details, see Tracey German, *Russia's Chechen War* (London: RoutledgeCurzon, 2003), 119–123.

27. German, 122.

28. Speaking at a press conference after hostilities began, Russian Deputy Minister of Foreign Affairs Grigory Karasin described the Russian action as a "peace-coercion operation." Transcript of Remarks and Response to

Media Questions by Russian Deputy Minister of Foreign Affairs/State Secretary Grigory Karasin at Press Conference at RIA Novosti News Agency, Moscow, August 10, 2008, https://www.mid.ru/brp_4.nsf/e7 8a48070f128a7b43256999005bcbb3/44465c679531114bc32574a300346fce? OpenDocument.

29. For further details on the Russian invasion, see Pavel Felgenhauer, "After August 7: The Escalation of the Russia Georgia War" in *The Guns of August 2008: Russia's war in Georgia*, eds. Svante E. Cornell and S. Frederick Starr (Armonk, NY: M. E. Sharpe, 2009), 162–180.

30. The Independent International Fact-Finding Mission on the Conflict in Georgia, *Report*, September 2009, 36, https://www.echr.coe.int/Documents/HUDOC_38263_08_Annexes_ENG.pdf.

31. Anton Lavrov, "Timeline of Russian-Georgian hostilities in August 2008" in *The Tanks of August*, ed. Ruslan Pukhov, (Moscow: Center for Analysis of Strategies and Technologies 2010), 43.

32. Civil.ge, "National Security Council Chief Testifies Before War Commission," October 28, 2008, https://civil.ge/archives/117793.

33. Quoted in Nathan Leites, "The Soviet Style of War" in *Soviet Military Thinking*, 216.

34. Leites, 216.

35. Oleg Kozhenko, "Kakim budet voennoe obrazovanie?," *Armeiskii sbornik* no. 3, (March 2009): 8.

36. Viktor Kutishchev, "Voinyi budushchego: kakimi im biyt?," *Armeiskii sbornik* no. 4 (April 2012): 63.

37. Kutishchev, 63.

38. For further details see Bettina Renz, *Russia's Military Revival* (Cambridge: Polity Press, 2018).

39. John Markoff, "Cyberspace becomes a new battleground," *International Herald Tribune*, August 13, 2008, 3.

40. Markoff, 3.

41. Dancho Danchev, "Coordinated Russia vs Georgia cyber attack in progress," *ZD Net*, August 11, 2008, https://www.zdnet.com/article/coordinated-russia-vs-georgia-cyber-attack-in-progress/.

42. N. A. Molchanov, "Informatsionnyi potentsial zarubezhnykh stran kak istochnik ugroz voennoi bezopasnosti RF," *Voennaya mysl'* no. 10 (October 2008): 2–8.

43. Andrew E. Kramer, "Georgia, outgunned, wages a war of words," *International Herald Tribune*, August 13, 2008, 1.

44. Molchanov, "Informatsionnyi potentsial zarubezhnykh stran kak istochnik ugroz voennoi bezopasnosti RF," 7–8.
45. Ministry of Foreign Affairs of the Russian Federation, "Concept of the Foreign Policy of the Russian Federation," approved by President of the Russian Federation Vladimir Putin, February 12, 2013.
46. "Vyistuplenie nachal'nika general'nogo shtaba vooruzhennykh cil rossiiskoi federatsii na temu: 'Voennyie opasnosti i voennyie ugrozy rossiiskoi federatsii v sovremennykh usloviyakh,'" *Armeiskii sbornik* no. 5 (May 2015): 58–63.
47. *Krym: put na rodinu*, directed by Andrei Kondrashev, released in March 2015.
48. *Krym: put na rodinu*.
49. For further details, see Jen Weedon, "Beyond 'Cyber War': Russia's Use of Strategic Cyber Espionage and Information Operations in Ukraine" in *Cyber War in Perspective: Russian Aggression in Ukraine*, ed. Kenneth Geers, (Tallinn: NATO CCD COE Publications, 2015) https://ccdcoe.org/uploads/2018/10/CyberWarinPerspective_full_book.pdf.
50. "Vyistupleniye General'nogo sekretarya ODKB NN Bordyuzhi na plenarnom zasedanii III Moskovskoi konferentsii po mezhdunarodnoi bezopasnosti" (speech), Moscow Conference on International Security, May 23, 2014, http://mil.ru/mcis-2014/multimedia/video/more.htm?id=5021@morfVideoAudioFile.
51. See for example Colm Quinn, "Russia Moves to Censor Domestic Media," *Foreign Policy*, March 2, 2022, https://foreignpolicy.com/2022/03/02/russia-media-ukraine-tvrain-echo/.
52. Anatoly D. Tsyganok, *Voina v Syrii i yee posledstviya dlya Blizhnego Vostoka, Kavkaza i Tsentral'noi Azii: russkii vzglyad* (Moscow: AIRO-XXI, 2016), 263.
53. "Segodnya vooruzhennyie silyi Rossii sposobnyi reshat zadachi luboi slozhnosti," *Armeiskii Sbornik* no. 12 (December 2017): 19.]
54. Mikhail Rostovskii, "Sergei Shoigu rasskazal kak spasal Rossiiskuyu Armiyu," *Armeiskii Sbornik* no. 11, (November 2019): 9.
55. Quoted in Vladykin, "Voennaya nauka smotrit v budushchee."
56. Gerasimov, "Po opytu Sirii."
57. Quoted in Vladykin, "Voennaya nauka smotrit v budushchee."
58. Dvornikov, "Formyi boevogo primeneniya i organizatsiya upravleniya integrirovannyimi gruppirovkami vooruzhennyikh sil na teatre voennyikh deistvii," 38.

59. Dvornikov, "Formyi boevogo primeneniya i organizatsiya upravleniya integrirovannyimi gruppirovkami vooruzhennyikh sil na teatre voennyikh deistvii," 38.
60. "Segodnya vooruzhennyie silyi Rossii sposobnyi reshat zadachi luboi slozhnosti," 18.
61. Ibid.
62. Ministry of Defense of the Russian Federation (website), "Russian Defence Minister General of the Army Sergei Shoigu holds teleconference with leadership of Armed Forces," February 5, 2019, https://eng.mil.ru/en/news_page/country/more.htm?id=12215894@egNews.
63. Oleg Falichev, "Robotyi vo glave c 'voevodoi,'" *Voenno-promyshlennyi kur'er* 706, no. 42, November 1, 2017, https://vpk-news.ru/articles/39639?.
64. "Segodnya vooruzhennyie silyi Rossii sposobnyi reshat zadachi luboi slozhnosti," 18.
65. Oleg Falichev, "Karuseli dlya protivnika," *Voenno-promyshlennyi kur'er* no. 32, August 25, 2020, https://vpk-news.ru/articles/58311.
66. Valery Stitzberg and Oleg Pochinyuk, "Svyazistyi ispol'zuyut siriiskii opyit," *Krasnaya Zvezda*, April 11, 2018, http://redstar.ru/svyazisty-ispolzuyut-sirijskij-opyt/.
67. Alexandr Tikhonov, "Siriiskaya proverka boem," *Krasnaya Zvezda*, August 29, 2017, http://archive.redstar.ru/index.php/component/k2/item/34260-.
68. Ministry of Defense of the Russian Federation (website), "Exchange of experience gained during operation in Syria is one of the goals of the operational session of the Russian Armed Forces leadership," July 10, 2017, https://eng.mil.ru/en/news_page/country/more.htm?id=12133809@egNews.
69. "V masshtabnom uchenii Vostok-2018 budut zadeystvovany osnovnyye sily tsentralnogo voennogo okruga," *Rambler.ru*, August 30, 2018, https://news.rambler.ru/middleeast/40685993-v-masshtabnom-uchenii-vostok-2018-budut-zadeystvovany-osnovnye-sily-tsentralnogo-voennogo-okruga/; Falichev, August 2020, https://vpk-news.ru/articles/58311.
70. Viktor Khudoleev, "Myi dokazali svoya boesposobnost," *Krasnaya Zvezda*, October 12, 2020, http://redstar.ru/my-dokazali-svoyu-boesposobnost/
71. Oleg Pochinyuk, "S uchyotom siriiskogo opyita," *Krasnaya Zvezda*, May 27, 2019, http://redstar.ru/s-uchyotom-sirijskogo-opyita-2/. For analysis of the role of UAVs in Russia's Syria campaign see O. V. Milenin and A. A. Sinikov, "O roli aviatsii vozdushno-kosmicheskikh sil v sovremennoi

voine. Bespilotnyie letatel'nyie apparaty kak tendentsiya razvitiya voennoi aviatsii," *Voennaya mysl'* no. 11 (November 2019): 50–57.

72. Oleg Pochinyuk, "S uchyotom siriiskogo opyita."
73. Tikhonov "Siriiskaya proverka boem."
74. NewsFront, "Intervyu nachalnika Genshtaba VS RF Gerasimova ob itogakh operatsii VS RF v Siriii i o dalneishikh perspektivakh siriiskoi voinyi," December 27, 2017, https://news-front.info/2017/12/27/intervyu-nachalnika-genshtaba-vs-rf-gerasimova-ob-itogah-operatsii-vs-rf-v-sirii-i-o-dalnejshih-perspektivah-sirijskoj-vojny/.
75. Gerasimov, "Razvitie voennoi strategii v sovremmenykh usloviyakh. Zadachi voennoi nauki," 6–11.
76. Dmitrii Semyenov, "V prioritete – siriiskii opyit," *Krasnaya Zvezda*, July 6, 2018, http://redstar.ru/v-prioritete-sirijskij-opyt/.
77. Gerasimov, "Razvitie voennoi strategii v sovremmenykh usloviyakh. Zadachi voennoi nauki," 9.

PART II

CONTINUITY AND CHANGE

CHAPTER 4

HIGH-TECH FUTURES

Prior to Russia's invasion of Ukraine in February 2022, Western analyses of Russian strategy tended to focus on concepts such as hybrid warfare and gray-zone operations, revealing a belief that Russia has directed its attention primarily toward nonmilitary means of achieving their strategic objectives. However, a significant amount of Russian military theoretical literature, as well as policy and procurement practice, has in reality centered around improvements to military means, both weapons and systems—in particular C4ISR systems (command, control, communications, computers, intelligence, surveillance, and reconnaissance)—and around enhancing the country's conventional military capabilities through the acquisition of high-tech assets and encouraging innovation in new, potentially disruptive technologies. The prevailing assumption that Russia has focused on asymmetric means because of an understanding that it cannot compete directly with US and allies does not provide the full picture: Russia has continued to focus on conventional military means, learning lessons from its own experiences, observing and emulating the US and allies, and seeking to overcome the perceived technological gap, while also making use of indirect, nonmilitary means, partly because it has concluded from its own observation that that is what states do in the contemporary era. The country's military-modernization program was driven partly by concern of a significant technological gap between Russia and Western states, particularly the US and its NATO allies. As discussed in chapter 2, Western military operations since 1991 have played a key role in the evolution of Russian military thought and the country's defense policy. The acquisition and use of modern high-precision weaponry by the US and its NATO allies was perceived to constitute one of the principal threats to Russian national security and prompted a shift in procurement focus.[1] Operational experience in Georgia confirmed the need for Russia to develop a precision-strike capability and improve its command and control systems. This was reflected in Gerasimov's 2013 piece in *VPK*, which sought to address the questions of what constituted modern warfare and what the armed forces should be prepared for and equipped with, and set out the defense establishment's view of the modern battlespace:

> The role of mobile groups of troops operating in a single reconnaissance and information space is increasing due to the use of new capabilities of control and support systems. New information technologies have significantly reduced the spatial, temporal and information gap between troops and control bodies...The use of high-precision weapons is becoming widespread. Weapons based on new physical principles and robotic systems are being actively introduced into military affairs.[2]

Putin emphasized the central role of advanced technologies in the modernization of Russia's armed forces, as well as its impact on contemporary conflict. Speaking at a series of meetings with the leadership of the Russian Ministry of Defense in November 2021, he stated that the development of production of "effective, innovative hi-tech weapons" was to be a growth area for the Russian armed forces. He described the development of laser, hypersonic, kinetic, and "other types of weapons" as a huge breakthrough in Russia's military technology, which has "significantly boosted the capacity of the Russian Armed Forces, ensuring a high level of Russian military security for many years, and even decades, to come and it helped strengthen our strategic parity."[3] This direction is reflected in the Russian State Armament Program covering the period 2024 until 2033, which focuses on high-precision weapons, including hypersonic weapons, the introduction of robotic systems, weapons based on new physical principles, electronic warfare equipment, and command and control systems based on artificial intelligence.[4] These plans are likely to be impeded by the sanctions regime imposed on Russia in the wake of its 2022 invasion of Ukraine, which has significantly limited access to key components such as semiconductors and microchips.

This chapter addresses how the Russian military thinking on changes in the character of conflict has focused on the increasing role of information and communications technologies, drawing particular attention to the enduring Russian focus on the military aspects of conflict. Western debates about the concept of network-centric warfare have directly influenced the direction of Russian military thought, leading to the development of the analogous Russian concept of network-centric warfare, which is considered to be both a key enabler and force multiplier. This chapter also examines the impact of technological advances and how they have been translated into specific action, examining some of the changes that were evident on the ground prior to the 2022 invasion of Ukraine. The results of these shifts in thinking at both the military and political levels, and subsequent modernization, have been evident in Russian military activity and weapons development such as the Kalibr missile, as well as initiatives such as the establishment of the NTsUO. Since 2015, Syria provided the Russian armed forces with a testing ground for new concepts and capabilities, and there has been growing use of precision strike and UAVs, as well as automated command and control. These changes reflect military transformations that have occurred around the world in the "information era": Russia is no different in this respect—it just came a little later to the party. The integration and application of new technologies, in particular information and communication technologies, to enhance the speed of decision-making, as well as the efficiency and effectiveness of military operations, underpins this transformation.

Efforts to improve the timeliness of decision-making reflects a Russian desire to hold the initiative and exploit the element of surprise, as discussed in chapter 1. Thus, the Russian military had been learning from its operational experience in Syria, after years of observation (and emulation) of Western approaches, assessing the implications for Russia and adapting accordingly. However, its failure to achieve a swift, decisive victory over Ukraine in February 2022 emphasized the perils of placing too much faith in technology at the expense of fundamentals such as troop morale, as well as the problems associated with drawing lessons from the observation of others (or your own experience). Precision strike did not have the intended effect during the opening days of the Russian invasion, raising questions about the extent to which some of the assumptions of Russian military theorists about how wars would be fought may have been flawed. Moscow appeared to be working on the assumption that long-range missile strikes and a large-scale invasion of ground forces would lead to a swift surrender by the Ukrainian government, underestimating the strength of Ukrainian resolve to resist and defend their homeland.

TECHNOLOGY AND FUTURE WARS

Russian military theorists have been reflecting on the significance of advances in technology for Russia and the potential impact on the character of conflict for decades; during the Soviet era, technological and scientific innovations such as nuclear weapons, space reconnaissance systems, air and missile defense systems, and automated command and control were deemed to be having a profound impact on military art.[5] The RMA mooted by Ogarkov in the 1980s was perceived to be driven by advanced technologies, particularly precision-strike and information technology. Vladimir I. Orlyansky was one of the first to analyze the impact of changing technologies on armed conflict in the twenty-first century. In a 2002 article he described the impact of the "informatization" of military affairs, emphasizing the importance of seizing and maintaining information superiority over an adversary in the buildup to, and during, military operations.[6] Khamzatov identified two contradictory dynamics shaping contemporary military operations: a reduction in the numerical size of national armed forces and the acquisition of high-tech systems and weapons, the latter of which has improved combat capabilities.[7] In 2004, Bogdanov set out his views of warfare of the future, asserting that in future wars, precision weapons, weapons based on new physical principles and information warfare would be prioritized.[8] Some analysts envisaged a high-tech future that would involve wars of the "seventh generation": in a 2009 piece on Russian military doctrine, Vorobyev and Kiselyev wrote about the "urgent need" for a rethinking of military transformation in light of the changing conditions. They urged the development of an asymmetric military policy in order to prepare the country and the armed forces for wars of the "seventh generation," which in their opinion included network-centric, cyber, information, space, robotic, non-contact, remote, nanotechnological, and intellectual wars (among other things).[9] They likened the impact of the "nanotechnological" era on military affairs to the transformation that followed the development of nuclear weapons, reasoning that there has been a major change in the character of conflict as a result of this. Technological superiority, achieved through the acquisition of modern precision-guided weaponry, was deemed to have replaced mass and numerical superiority in terms of troops and weapons as the key to defeating an adversary, and it was vital for the Russian armed forces to "catch up" (assumptions that have been undermined by events in Ukraine since 2022).[10] The pair built on this in 2011, emphasizing the rapid pace of scientific and technological progress, which had resulted in new types of weapons and the automation of command and control.[11]

A 2021 analysis argued that the desire of states to "neutralize" an adversary's nuclear advantage had stimulated the development of more powerful conventional weapons and systems that could deliver a preemptive strike. This was perceived to be influencing the character of conflict and war, with war shifting from Slipchenko's fifth generation to the sixth. Key areas for development included hypersonic missiles, weapons based on "new physical principles," high-precision long-range missiles, military robotics, UAVs, and AI in weapons and military equipment.[12] This reflects a common view that twenty-first-century warfare would be dominated by precision-strike, the critical importance of information superiority, the widespread use of UAVs for a range of tasks including to break through air defense systems, deployment of robotic systems and elements of AI, and the use of military and nonmilitary means.[13] Chekinov and Bogdanov argued that the era of classical war had come to an end, replaced by "wars of a new technological era," an era of "computer science, space, electronics, robotics, network-centric control, artificial intelligence," increasing the role and importance of military forecasting and foresight.[14] The evolution of the war in Ukraine, following Russia's failure to achieve a swift victory in February 2022, challenged these assumptions around the critical role of technology and the myth of perpetual progress. Russian forces abandoned precision strike and switched to an approach that they have resorted to many times since 1991 against cities such as Grozny in Chechnya and Aleppo and Idlib in Syria: the use of heavy, indiscriminate artillery and aerial bombardments to destroy urban areas, imposing heavy costs on the Ukrainian population and their leaders.

Military science is deemed to have a central role in the development of modern armed forces, with some arguing that it is impossible to modernize without the input of military science. Saifetdinov has suggested that it is impossible to effectively manage armed forces in the contemporary era without an automated control system, which "meets modern requirements arising from the study of the essence and content of the nature of a future war."[15] Kopytko and Kopylov have also emphasized the key role of military science in the development of the armed forces, condemning the failure of previous governments to implement proposed reforms, which in their view contributed to the ongoing technological gap with the West. They are very critical of the outflow of scientists, specialists, and engineers who have left Russia for the West; in their opinion, Western states have benefited from their expertise and have used it to assist in the creation of cutting-edge weaponry before reforming their armed forces, unlike Russia, which they accuse of putting "the cart before the horse."[16]

As set out in chapter 2, there has been considerable monitoring of the operational experiences of foreign armed forces, particularly Western ones, with the apparent intention of learning lessons that could be useful for the Russian military, adapting and emulating where necessary. This observation reinforced concerns that Russia was lagging behind other states, particularly in terms of modern technology: in 2013, Anatoly Kulikov, a former Army general, expressed his belief that Russia was lagging at least fifteen years behind the West.[17] This echoed V. Kutyishchev's 2012 analysis of future war, which focused predominantly on the impact of NCW on Russian thinking about conflict and future war. He argued that the Russian armed forces were still focused on outdated views of traditional, large-scale ground operations, rather than "modern concepts that involve the widespread use of precision weapons," emphasizing that the creation of a truly integrated group of forces, aided by good communications and coordination, was an "essential condition" for conducting operations in accordance with the principles of network-centric warfare.[18] Thus, there is a particular focus on the central role of a network-centric approach in twenty-first-century warfare.

The concept of network-centric warfare (or network-centric operations[19]) was developed in the US in the 1990s,[20] driven by the emergence of modern information and communications technologies that facilitated the development of networked systems of sensors, command and control systems, and weapons platforms. The concept was central to the transformation of militaries around the world in pursuit of improved operational efficiency and effectiveness. Information and communication technology advances transformed the way in which forces could communicate, ensuring they (and political decision-makers) had access to up-to-date information, intelligence, and situational awareness. A number of states developed their own definitions of the concept and adapted their armed forces accordingly, including France, the UK, and

China. Russian interest in NCW, which began to be widely debated in the early 2000s, thus stemmed from international developments and concerns that it was being left behind. As discussed in chapter 3, Russia's own experience of military conflict during the post-Soviet era has had a significant influence on the ongoing military transformation and the shift toward high-tech platforms and equipment. One of the principal problems experienced during the wars in Chechnya in the 1990s and 2000s was the challenge of coordination and communication: Russia's experience emphasized the need for secure, modern command, control, and communications systems to ensure that different force structures could coordinate. These issues had not been resolved by the time of Russia's intervention in Georgia in 2008, when one of the key problems identified was obsolete communications systems and poor land/air integration. Russia's military modernisation program, initiated in the wake of the 2008 war with Georgia, began to focus on the introduction of C4ISR, the development of credible network-centric capabilities for Russia, and the introduction of high-precision weapons.

NETWORK-CENTRIC WARFARE AND CAPABILITIES

The concept of network-centric warfare and analysis of the supposed network-centric wars of the West has been a particular focus of much of the military theoretical work in Russia since the early 2000s.[21] The concept has been widely debated in Russian military journals, as military scientists and experts analyze its impact on the character of conflict and the implications for Russia. Considerable attention has been devoted to Western approaches to network-centric warfare, with experts examining both the West's theoretical literature and operational experience in order to draw lessons for Russia. Based on evidence from operations in Afghanistan and Iraq, the modernization of US forces to incorporate network-centric capabilities is perceived to have greatly improved their efficiency and control.[22] In 2011, Kiselyev and Vorobyev, among others, drew attention to NCW, as practiced by the US and its NATO allies, and expressed concern that Russia was being left behind and needed to catch up. They called for the improvement of information management systems to be made a priority in order to create a secure, integrated, and resilient "information network with a single communications structure that ensures close interaction between information systems and intelligence services, as well as real-time decision-making and control of weapons systems." Improvements in control systems at the strategic, operational, and tactical levels would facilitate a transition to "new forms and methods" for managing modern combat systems, including means of reconnaissance and strike, electronic warfare, robotic military equipment, and UAVs, as well as the integration of forces and means of control.[23]

Nikolai N. Tyutyunnikov provided one of the clearest definitions of NCW, from a Russian perspective, defining it as a "concept of military operations oriented towards achieving information superiority that boosts the combat power of...integrated forces through the creation of an information and communication network linking sensors (data sources), decisionmakers and assets, which ensures that the participants of operations have situational awareness, accelerating command and control as well as increasing the tempo of operations, effectiveness of defeating enemy forces, survivability of troops, and level of synchronisation."[24] Bogdan I. Kazar'yan defined it as "operations and combat actions, in which the interaction of forces, command and control, and reconnaissance, is facilitated by networked means" and sets out a number of prerequisites for NCW.[25] A. Usikov identified the basis of network-centric capabilities as the creation and deployment of "automated electronic networks of intelligence, information and control in a theatre of operations." These networks are interconnected and linked into a single information and management complex, providing continuous, flexible and stable control of forces, as well as weapons systems.[26] Kiselyev and Vorobyev note that, although the concept of network-centric war was being used more frequently in Russia, usually in reference to Western operations, the term "conduct of hostilities in a single information and communication space" was used.[27] Chekinov and Bogdanov examined the concept of network-centric warfare in their work on "new-generation warfare," defining it as a concept of command and control that reflected a new way of directing armed forces in the twenty-first century. In their view, network-centric warfare had evolved as a result of the rapid development of IT and the creation of high-precision weapons and weapons based on new physical principles.[28]

According to Russian military scientists, one of the principal advantages of NCW is the high manuverability it lends forces in theater, underpinned by a belief that troops equipped with IT have a combat potential three times higher than conventional units.[29] This has fostered a conviction that information superiority will enable better, more rapid decision-making, giving an actor the advantage over an adversary, and facilitate parallel, continuous operations. Khamzatov has described NCW as "blitzkrieg" for a new generation, facilitating the seizure of the strategic initiative during the initial period of war and preventing an adversary from responding effectively by undermining its ability to deploy.[30] Others have been more circumspect in their analysis, suggesting that NCW is not a concept setting out how to conduct war in the twenty-first century; instead, they assert that should be understood as a management concept, which reflects a new means of directing armed forces.[31] Kazar'yan agreed with this assessment, emphasising that NCW is not a form of warfare but a characteristic of it, defined by the widespread use of automated, networked systems. He questions the utility of the term "network-centric," arguing that it makes no more sense than artificially constructed phrases such as "scattered-centralised" or "flat-convex"; in his view, "the brain of living organisms, systems of group behaviour...have nothing in common with network-centrism."[32] Nevertheless, he provides a wide-ranging analysis of the concept and its significance for Russia, cautioning that, although NCW accentuates the overwhelming technical superiority of the US armed forces, it has not been developed on the basis that its adversaries are weak and assumes others will catch up.

There is agreement among Russian experts that the implementation of methods associated with NCW has prompted an evolution in the forms and methods of warfare, demonstrated clearly by the US-led operations in Iraq in 1991 and 2003, the opening phase of which is deemed to have been characterized by surprise and unpredictability.[33] NCW is seen as a force multiplier that can enhance military capabilities and thus can hopefully close the gap with NATO militaries. Although the Russian focus on NCW has been driven by concern about the network-centric capabilities of its competitors (both in the West and China[34]) and fears that Russia is lagging behind in terms of its military technology, some have warned about falling into the mental trap of an arms race with the West, arguing that "network-centrism" is not a panacea for all the shortcomings of the Russian armed forces; the shortcomings of NCW need to be assessed alongside the challenges of implementing it. They point out that Western NCW operations have, to date, been conducted against weaker adversaries who were not equipped with satellite reconnaissance, precision-guided weapons, or automated command and control—their implicit message is that even if Russia perfects NCW capabilities, it will not necessarily gain any military advantage over the US and its allies.[35] There has also been discussion of the vulnerabilities of networked systems and the importance of developing further defenses against electronic warfare such as the jamming of reconnaissance assets.

These debates have not remained in the abstract realm of theory; there have been significant changes to the process of military decision-making, as well as weapons systems, that exploit technological advances and reflect the conclusions of military scientists and experts outlined previously. Key changes include the digitization of decision-making equipment, the introduction of automated command and control systems, and the development of weapons such as precision-guided munitions and hypersonic missiles. The establishment of the NTsUO is representative of the transformation. This was put into operation in December 2014, representing a significant step forward in the integration of Russia's Defense and security structures, both at the strategic and operational level, as well as the digitization of its processes: the Defense Minister Sergei Shoigu boasted that the Center is equipped with "the most powerful computer systems," alongside advanced communications technologies and automated systems.[36] According to the Ministry of Defense, it is intended to centralize command and control of the Armed Forces; oversee the day-to-day management of the military; and collect, compile, and analyze information on the military-political situation around the world, as well as on strategic direction and the sociopolitical situation in Russia during both peace and war. It provides a single digital platform, which unites information flows from the Armed Forces, federal executive bodies, and state corporations such as Rosatom.[37] The NTsUO comprises three components: the Strategic Nuclear Forces Control Center; the Combat Control Center, which manages the rest of the Armed Forces and other security agencies; and the Daily Operations Control Center, which coordinates logistics and civil agencies. It is designed to take a whole-of-government approach and brings together a number of defense and security bodies, including the Ministry of Defense, General Staff, the Ministry of Internal Affairs, Ministry of Emergency Situations' Crisis Center, and Rosatom (the national nuclear corporation), among others. The establishment of the NTsUO has been described as a significant step toward improved efficiency across Russian defense as a whole, not just the armed forces, and Gerasimov has credited it with "radically" changing the Russian approach to managing the military, evidenced by operations in Syria, which is viewed as a successful example of a whole-of-government approach.[38]

The fifth anniversary of the NTsUO was marked in 2019 with a conference at the Military Academy of the General Staff, which focused on the results of the Center's operation as well as future developments. The Center's head, Colonel General Mikhail Mizintsev, commended the creation of the digital platform that integrated information from various government and military bodies, noting that the ongoing adaptation of military command and control was essential in the contemporary information age. He also accentuated the critical role of military science in the development of an understanding of the implications of technological innovations for the armed forces and Russian security more broadly.[39] Saifetdinov echoed this view in 2020, drawing a direct link between the enduring utility of military science and automated command and control systems for the armed forces. He argued that it was impossible to create an effective automated command and control system for the Russian armed forces without clarification of the operational requirements of such a system, resulting from the study of the character of future wars.[40]

The NTsUO was viewed as an important step in the creation of a unified information space for the armed forces, the development of which has been underway for a number of years, prompting widespread debate.[41] Gerasimov has referenced this, stating in 2018 that delivery of automated control systems for troops and weapons would begin that year as the development of "modern means of combat control and communication, integrated into a unified information space, is underway."[42] Progress has been slow, however, and a number of problems have been identified of an organizational, technological, and technical nature. According to analysis by Anatoly Ya. Chernysh and Vladimir V. Popov, problems include the lack of a single coordination center to manage the creation of a unified information space; the lack of agreed methodology; the wide range of different IT systems in use; and no agreement about which tools to use for data processing.[43] Nevertheless, in spite of slow progress in terms of the creation of a unified information space for the Russian armed forces, overall the development of a network-centric capability had advanced considerably.

Electronic Warfare: The Other Side of the NCW Coin?

The Russian focus on network-centric approaches has accentuated the vulnerabilities of networked systems, prompting a renewed focus on electronic warfare during the first decade of the twenty-first century. As military theorist N. I. Sidnyaev put it, "the whole concept of a network-centric system loses its meaning if protection from electronic warfare is not provided."[44] Electronic warfare is considered to be a central characteristic of future war, and a number of military theorists have covered the topic, linking electronic warfare with the criticality of gaining information superiority during the initial period of war. Slipchenko characterized the forces and means of electronic warfare as a key component of "sixth-generation" contactless wars, along with precision strike and weapons based on new physical principles (to be discussed in the following paragraphs).[45] In 2015, Yury E. Donskov, a retired colonel and leading researcher at the Institute of Electronic Warfare, published a number of articles on electronic warfare and the implications for Russia. He asserted that in the future (until 2030) the forces and means of electronic warfare should be granted priority in air defense and the protection of infrastructure from aerial bombardment.[46] In another article, he analyzed electronic warfare within the context of NCW and argued that changes in the communications systems used by the militaries of "leading foreign states" necessitated "new approaches to the construction of a spatially distributed electronic warfare system" that took into account technological advances as well as the nature of modern military operations.[47]

Yury E. Donskov, V. I. Zimarin, and B. V. Illarionov argue that EW (electronic warfare) troops are an essential component of the Russian Ground Forces, deployed to "disorganize" the information support for an adversary's operations, while protecting Russian troops and assets from precision strike. In the twenty-first century, electronic warfare is thus considered to be on the same level as infantry, armor, missiles and artillery, and air defense, tasked with gaining information superiority over an adversary through disorganization of their information support.[48] Disorganizing an adversary's command and control systems—that is, actions to disrupt and erode command and control systems to undermine control of forces and achieve information superiority—is central to Russian military thinking, and is an area where electronic warfare is critical. There has been a considerable focus on electronic warfare in Russian military thinking over the past decade or so, alongside significant investment in EW capabilities as part of the post-2008 military modernization process.

Sergei Shoigu claimed in 2021 that many Russian EW systems were unparalleled and superior to foreign systems and had performed well during testing in Syria.[49]

HIGH-PRECISION WEAPONS

Precision strike has been another area of significant debate and analysis within the Russian strategic community. Slipchenko's concept of "sixth-generation" warfare, discussed in chapter 2, emphasizes the decisive role of high precision conventional weapons (both offensive and defensive) and the secondary role of land forces, so-called "contactless or non-contact war," which aims to decimate an enemy's economic potential and instate a change of political regime. A number of analysts have advanced this idea, arguing that precision-guided long-range weapons were changing the character of future conflict. Vladimir A. Vinogradov identified a shift toward non-contact operations and the use of high-precision weapons, noting the central role of surprise in such operations, together with air supremacy, while Gareev warned that the impact of precision-guided munitions should not be underestimated, stating that in future "wars will be contactless," and there would be no boots on the ground.[50]

Vorobyev and Kiselyev believe that high-precision weapons, together with new means of electronic warfare and space-based reconnaissance and guidance systems, comprise the principal drivers of the changing character of conflict. In their opinion, the very art of war has been raised to a new level by the qualitative change in weapons systems, reconnaissance, electronic warfare, and automated command and control systems that has occurred since the 1990s. They called for the urgent development within Russian military science of a theoretical basis for high-precision battles, which they saw as replacing traditional methods of armed confrontation, arguing that "precision combat" is a particular form of armed struggle, pointing to evidence from the 1991 Gulf War, Operation Allied Force, and the 2003 Iraq War, where precision strikes are deemed to have achieved key objectives including the defeat of enemy forces, the realization of air superiority and supremacy, the disorganization of the enemy's political and military command, as well as the destruction of critical infrastructure.[51] Key features of high-precision operations, according to Bogdanov and Kiselyev, include the multifaceted nature of the beginning of hostilities and the role of special forces, who are often in theater long before any strike is launched, for reconnaissance purposes, as well as sabotage and subversion. Not all Russian theorists agreed with the central role of high-precision weapons in contactless war. Chekinov and Bogdanov agreed that precision strike had changed the character of contemporary conflict and should not be underestimated but warned that it had not rendered other advanced forms and methods of "contact" war meaningless. According to their analysis, high-precision weapons would play a central role in future war, alongside UAVs and autonomous weapons.[52]

The abundant use of high precision weapons was considered to be as effective at the strategic, operational, and tactical levels as nuclear weapons. Although US missile defense was perceived to have undermined the importance of nuclear weapons, eliminating Russia's strategic parity, the debate among Russian experts suggests that they believe high-precision weapons and hypersonic missiles could go some way to restoring this parity. Equipping the Russian armed forces with high-precision and hypersonic weapons systems was believed to considerably boost its conventional military capabilities and was expected to pose a long-term challenge to potential adversaries. Furthermore, the use of high-precision long-range missile strikes is today deemed by some theorists to be an effective means of preempting an adversary and therefore ensuring that Russia has the initiative.[53] Zarudnitsky has emphasized that victory in future wars will be dependent on an actor achieving superiority over an adversary and gaining (and retaining) the initiative, stressing that in order to achieve this the development of hypersonic missiles will remain a priority direction for Russia. He asserts that contemporary Russian military strategy involves a new approach to the integrated use of all available forces and means to defeat an adversary, facilitating the achievement of superiority in firepower: "the comprehensive defeat of the enemy will be achieved by advance planning of all types of impact, providing a phased transition from strategic deterrence to direct fire."[54] Stepshin and Anikonov have suggested that the use of precision strike enables a strategy of "selective action," whereby surgical strikes are directed against critical targets, inflicting high costs against an adversary at a low cost for the protagonist.[55] These assumptions were undermined by the initial failure of the Russian military in Ukraine in 2022: long-range missile strikes and a multipronged invasion of ground forces did not lead to a swift surrender by the Ukrainian government, and Russia quickly lost the initiative, having underestimated the strength of Ukrainian resolve to resist and defend their homeland. Subsequently, Russian forces returned to their traditional modus operandi, the use of overwhelming, indiscriminate military firepower, particularly Bellamy's so-called "God of War" artillery, leading to widespread destruction and civilian casualties.

In their 2006 analysis of the impact of precision strike on the character of conflict, Vorobyev and Kiselyev emphasized the critical role of reconnaissance-fire (*razvedivatel'no-ognovoi kompleks*) and reconnaissance-strike complexes (*razvedivatel'no-udarnyi kompleks*).[56] The Russian military experience in Syria reconfirmed the central role of these concepts, which were initially developed during the Soviet era. The term "reconnaissance-strike complex" refers to the coordinated use of high-precision, long-range weapons linked together with intelligence, surveillance, and reconnaissance capabilities, as well as command and control, at the strategic level, while reconnaissance-fire is used at the tactical-operational level. Oleg V. Tikhanychev has noted that the compression of time on the contemporary battlefield has accentuated the need for rapid decision-making, facilitated by automated command and control systems. In his view, reconnaissance-fire and reconnaissance-strike complexes offer a solution, reducing what he terms the engagement cycle and optimizing the process of reconnaissance, decision-making and the use of weapons, thus enhancing military effectiveness.[57] Vorobyev has sought to apply the concept of reconnaissance fire and strike to information operations, arguing that the era of information confrontation has made the classical triad of "fire, strike, maneuver" irrelevant. He identified a new form of warfare, "electronic fire and information-strike" operations, adapting the concepts of reconnaissance-fire and reconnaissance-strike to the information realms and drawing attention to the key role of space and space-based assets.[58]

Russia's involvement in Syria reconfirmed to political and military leaders that the development of precision-guided munitions should remain a priority: the deployment of Kalibr missile strikes in Syria in October 2015 demonstrated the evolution of Russian thought and new capabilities in precision strike.[59] A central element of this is the development of hypersonic cruise missiles, which are able to circumvent advanced air-defense systems. The development of hypersonic missiles appears to have been driven largely by the US missile defense program, as well as its development of Prompt Global Strike, which is reported to be capable of delivering precision-guided conventional airstrikes anywhere in the world within an hour, encouraging the Russians to develop a similar capability. In his annual address to the Federal Assembly in February 2019, Putin accused the US of pursuing "absolute military superiority" with their missile defense plans and warned that the Russian response would be "efficient and effective."[60] As discussed in chapter 2, a long-running Russian concern regarding US missile defense centers around the loss of strategic parity with the US. Hypersonic missiles are seen as one way to regain parity. Speaking in 2019, Putin set out details of the Avangard hypersonic boost-glide vehicle, which was put into service in late 2019, and the Tsirkon hypersonic missile, saying it could strike targets over 1,000 kilometers away and reach speeds of Mach 9, and warned Russia's enemies to "calculate the range and speed" of the country's future arms systems.[61] The Tsirkon hypersonic missile, which is expected to come into service in 2022, is naval surface and sub-surface, and can be used by any surface naval vessel that is equipped for the Kalibr, although there are plans to develop a land-based version.[62] A number of successful test firings of the Tsirkon missile took place from the *Admiral Gorshkov* frigate during 2020 (including one in early October 2020, close to Putin's birthday, reflecting the date that the first long-range strike of the Kalibr missile from the Caspian into Syria took place in 2015). In October 2021, the Russian Ministry of Defense announced that a Tsirkon missile had been successfully launched from a submarine for the first time.[63]

LOOKING FORWARD: ARTIFICIAL INTELLIGENCE AND AUTONOMOUS WEAPONS

Putin held a series of meetings with the leadership of the Russian Ministry of Defense in late 2021, which covered a range of issues related to the integration of modern technology in the armed forces, including UAVS (as well as unmanned underwater vehicles, UUVs), AI, and weapons based on new physical principles. According to the president, the Russian armed forces, which have over two thousand UAVs, will continue to procure and develop them with "the use of artificial intelligence and the most advanced achievements of technology and science."[64] UAVs were used increasingly in Russia's Syria operation, and the armed forces gained valuable experience about integrating them into their force structure. According to Gerasimov, Russian forces were operating as many as seventy UAVs each day in Syria for reconnaissance and electronic warfare tasks (including reconnaissance-strike and reconnaissance-fire "contours"), prompting him to state that it was now "inconceivable to conduct hostilities without a drone."[65] In 2018 the Russian CGS referenced the development of multipurpose UAVs that would enable both reconnaissance and strike missions, part of improvements to the overall effectiveness of the strike capabilities of the armed forces (although at the time, armed UAVs were not being used in Syria). He also pointed to the creation of UAV units within military formations throughout the Russian armed forces in order to increase combat capabilities and ensure autonomy.[66] Gerasimov tied this directly to the changing character of conflict, stating that the contours of the "most likely future war" meant that counteracting communications, reconnaissance, and navigation systems through technologies such as UAVs would play a key role in future conflict. General Aleksandr Zhuravlev has said that Russian troops have been training with UAVs in military exercises, incorporating experience acquired during the operation in Syria to practice integrated operations involving ground forces working in cooperation with aviation and UAVs.[67] Russian thinking on UAVs is likely to have been reinforced by the important role they have played in the war in Ukraine: the Ukrainians made very effective use of Turkish TB2 Bayraktar UAVs to target Russian forces, the same UAVs used by Azerbaijan in Nagorno-Karabakh in 2020.

As a result of both its own operational experience in Syria and observation of the experience of others, discussed in previous chapters, Russian experts have become convinced that UAVs will play a central role in future war.[68] The Second Nagorno-Karabakh War reinforced the importance of UAVs in contemporary conflict, described in one analysis as the "era of remotely piloted aviation."[69] Consequently, UAV technologies will remain a key area of expansion moving forwards, in particular the development of a reconnaissance and strike capability to enable the increased use of UAVs for precision strike. A key element of this will be the S-70 Okhotnik, which is equipped with stealth technology and has been undergoing active tests since 2020.[70] The belief that UAVs will play a central role in future war has also led to consideration of how to counter the growing threat from UAVs, with V. V. Repin arguing that electronic warfare will be the primary tool to counter UAVs.[71]

The third meeting in the series of November 2021 meetings was devoted entirely to discussion of equipping the armed forces with weapons based on "new physical principles" (*oruzhiye na novykh fizicheskikh printsipa*). This Russian phrase refers broadly to weapons and military systems that operate on principles (either natural phenomena or physical processes) previously not used within the military realm. There has been significant coverage of weapons of new physical principles in the military thought literature, and a number of Russian military theorists have written about future wars being distinguished by the use of weapons designed on new physical principles, which will include geophysical, infrasonic, climate, laser, ozone, radiological, accelerator (beam), electromagnetic, directed energy, and non-lethal (such as psychotropic preparations and infrasonic weapons), as well as unconventional arms that might cause earthquakes, typhoons, or flooding that would lead to the erosion of economies and intensification of tension in an adversary's population.

AI also features prominently in Russian military thinking about future war and the development of the country's armed forces, reflecting official priorities as well as observation of the US (and increasingly China).[72] In 2019, Putin called for Russia to be a world leader in AI, describing Russian leadership on the issue to be of the utmost importance.[73] In the same year he warned that the development of AI carried both risks and opportunities, stating that if an actor gained a monopoly in the field of AI, they would "become the ruler of the world."[74] He reiterated this theme two years later at an international conference on AI held in Moscow in late 2021:

> The winners in today's world are those who are making better use of the powerful technological potential in the interests of people and their prosperity. They are winning the global competition. We, that is, Russia must certainly be among the leaders in this regard.[75]

Consequently, Russian military theorists have increasingly focused on the implications of AI for the character of conflict, and there have been a growing number of articles on the topic in military theoretical journals. This accentuates a widespread conviction that technological advances shape the contours of future war and conflict more than geopolitical circumstance. D. V. Galkin, P. A. Kolyandra, and A. V. Stepanov identified AI as one of the most important technologies to increase the potential of the Russian armed forces.[76] This has been echoed by Zarudnitsky, who has expressed his belief that AI will play a significant role in the future development of the armed forces, including AI that is capable of "self-learning and analysis of big data for applications...from intelligence and weapons management to strategic forecasting and decision-making":

> The rapid development of both military and nonmilitary means of confrontation, primarily with the use of artificial intelligence technologies, is facilitating the emergence of promising forms of employment of the Russian armed forces, from a strategic operation and an operation of strategic [deterrence] to a global military campaign.[77]

Vasily M. Burenok[78], an expert on weapons systems, has written widely on the implications of technological change for the Russian armed forces, focusing in recent years on the impact of AI. He set out his views on AI and future war in a 2021 article in *Voennaya Mysl*, asserting that "the creation and development of AI systems is currently becoming one of the most important areas of scientific and technological progress, the very fundamental technology that can radically change the nature of not only armed struggle, but also the whole essence of power confrontation between states, including economic, information and cyber war. This change will be characterized by the priority role of AI systems during this confrontation."[79] Key areas that he believes will see the introduction of AI technologies moving forwards include systems for processing and integrating information and intelligence data; control systems for collective action by robotic, troop, and mixed groups of weapons/military equipment; and targeting systems. In his view, one of the most obvious uses of AI in the near future will be to control drone swarms, which can be used to destroy or paralyze existing high-tech platforms and systems. Nevertheless, in spite of his conviction that AI will play a leading role in future wars, Burenok cautioned against overoptimism regarding AI capabilities, especially in terms of automated command and control systems.

Like other aspects of technological development, the focus on AI and robotics has not remained in the theoretical realm; over the past decade a number of official bodies have been created to oversee future developments. In 2012, Russia established the Advanced Research Foundation (*Fond perspektivnykh issledovanii*), mirroring the US Defense Advanced

Research Project Agency, to focus on AI, hypersonics, and other emerging and disruptive technologies.[80] The following year, in 2013, a Robotics Center was established under the aegis of the Ministry of Defense to develop military robotics systems, with marine robotics and UUVs being a particular focus.[81] Military research and development received a further boost with the creation of ERA Technopolis, a "military innovation city" located in Anapa on Russia's Black Sea coast. This was inaugurated by a 2018 presidential decree and includes research into a range of priority areas including AI, robotics, small spacecraft, automated control, and IT systems, informatics and computer engineering, nanotechnology, biotechnology and weapons based on new physical principles.[82] ERA is managed by the Russian MoD's Directorate for Innovative Development (*Glavnoye upravlenie innovatsionnogo razvitiya*) and at the end of 2021, the government announced that a department for the development of AI would be set up in 2022 (although this is likely to be impacted by the war in Ukraine).[83] Shoigu has announced plans to create Russia's first robotic military strike unit (in which humans will play a secondary role) capable of operating on a nuclear battlefield, while robots have been used for mine clearance in Syria and Nagorno-Karabakh.[84] Robotic units were deployed alongside troops for the first time during the Zapad-2021 military exercises, with the Uran-9 and Nerekhta reconnaissance and fire support robots engaging directly in operations.[85] There are still a number of technical challenges related to the further development of a military robotic capability within the Russian military, including their vulnerability to EW, but the use of robotic units on the battlefield of a future conflict is a long-term priority for the development of the armed forces, driven by the desire to minimize troop casualties (an intention also reflected in the increasing use of private military companies, discussed in chapter 6). Experts believe that the capabilities of robotic complexes (*robototekhnicheskiye kompleksyi*) will be significantly expanded by the creation of "multifunctional unified platforms" as a result of universal digitalization and the production of high-tech materials.[86]

Much Russian military theoretical literature in the post–Cold War era has been focused on the implications of technological development and advanced technologies both for the character of conflict generally and specifically for Russia. Moving forward, the Russians are already exploring AI and the role that emerging technologies can play on the battlefield. Network-centric warfare, which has been analyzed for years by the Russian military, is deemed to constitute a significant development in the means of warfare in the twenty-first century, prompting a transition from seeking the physical destruction of an adversary to the pursuit of a more complex influence, achieved by a single integrated system that includes precision strike, reconnaissance, electronic warfare, and information warfare with strategic, operational and tactical effect. Russia has been using technology to augment and improve existing military capabilities, and there have been considerable efforts to integrate and apply new technologies, particularly information and communications technologies, to enhance the speed of decision-making and improve the efficiency and efficacy of military operations, reflecting a desire to seize the initiative and take advantage of the element of surprise. Thus, the battlefield has not been changed by the incorporation of modern technological advances, but it has been improved; there has been evolution rather than revolution. Russia's desire to achieve "superiority of management" in terms of its command and control will necessitate the disruption of an adversary's command and control systems, using electronic and information warfare, as well as kinetic strikes. Slipchenko argued that "seventh-generation" warfare would be achieved by 2050, by elevating information systems to the level of a combat arm, not just combat support.

The enduring belief that technology remains a key determinant of how war is fought is central to Russian military thought, as evidenced by the focus on network-centric warfare, precision-strike, and information technologies. At first glance, Russia's initial invasion of Ukraine in February 2022 seemed to suggest a change in the approach of the Russian armed forces that incorporated ideas about the central role of precision-strike in contemporary warfare. The Russian political and military leadership was perhaps overly confident in the utility of conventional high-precision weapons to achieve success during its initial invasion of Ukraine in 2022, which witnessed significant use of Russia's precision-strike capabilities, such as the Iskander missile. However, precision strike did not have the intended effect during the opening days of the Russian invasion. Moscow appeared to be working on the assumption that missile strikes and a large-scale invasion of ground forces would prompt a rapid Ukrainian capitulation, underestimating the strength of Ukrainian resolve to resist and defend their homeland in the face of the Russian assault. The war in Ukraine has pushed against the enduring belief that technology remains a principal determinant of how war is fought, while also demonstrating the criticality of intangible factors such as morale and the will to fight and resist. Russia's plans for the development of high-tech platforms and weapons systems are likely to be impacted by the sanctions regime imposed on Russia in the wake of its 2022 invasion of Ukraine, which has significantly limited access to key components such as semiconductors and microchips. This could see it relying more on "low-tech" equipment, bolstering enduring anxiety about technological inferiority. Concerns about technological inferiority are a leitmotif in Russian military thought, particularly vis-à-vis parity with the US. This has driven a focus on the nonmilitary means of achieving strategic objectives through the targeting of an adversary's vulnerabilities, such as societal cohesion, in order to undermine its will to resist. The next chapter examines Russian thinking on information confrontation and how an adversary can be undermined from within, without necessarily having to resort to military intervention.

NOTES

1. Litvinenko and Yastrebov, "VTO: vzglyad v budushchee."
2. Gerasimov, "Tsennost' nauki v predvidenii."
3. President of Russia (website), "Meeting with Defence Ministry leadership and heads of defence industry enterprises," November 3, 2021, http://en.kremlin.ru/events/president/transcripts/67061.
4. Interview with Yury Borisov, *Interfax*, May 9, 2021, https://www.interfax.ru/interview/764864.
5. For example, see I. A. Lomov, ed., *Nauchno-tekhnicheskiy progress i revolyutsiya v voennom dele* (Moscow: Voenizdat, 1973).
6. Vladimir I. Orlyansky, "Vooruzhennaya i informatsionnaya bor'ba: sushchnost i vzaimosvyaz ponyatii i yavelenii," *Voennaya mysl'* no. 6 (November 2002): 42–43.
7. Musa M. Khamzatov, "Vliyanie kontseptsii setetsentricheskoi voiny na kharakter sovremennykh operatsii," *Voennaya mysl'* no. 7 (July 2006): 13–17.
8. Sergei A. Bogdanov, "O structure i soderzhanii voennoi nauki na sovremennom etape razvitiya voennoi mysli," *Voennaya mysl'* no. 5 (May 2004): 19–28.
9. Ivan Vorobyev and Valery A. Kiselyev, "Voennaya doktrina Rossii XX1 veka," *Armeiskii Sbornik* no. 1 (January 2009): 9.
10. Vorobyev and Kiselyev, 9.
11. They went on to argue that the growing role of IT, software modeling, and other networked activities necessitated a major change in professional military education. I. N. Vorobyev & V. A. Kiselev, "The role of military science in shaping the new look of the Russian Armed Forces," *Voennaya mysl'* 2 (February 2011): 40–48.
12. Stepshin and Anikonov, "Razvitiye vooruzheniya, voennoi i spetsial'noi tekhniki i ikh vliyaniye na kharakter budushchikh voin."
13. Stepshin and Anikonov, "Razvitiye vooruzheniya, voennoi i spetsial'noi tekhniki i ikh vliyaniye na kharakter budushchikh voin."
14. Chekinov and Bogdanov, "Voennaya futurologiya: zarozhdeniye, razvitie, rol i mesto v sisteme voennoi nauki," 29.
15. Kharis I. Saifetdinov, "Rol' voennoi nauki v sozdanii i razvitii avtomatizirovannoi sistemyi upravleniya Vooruzhennymi Silami Rossiiskoi Federatsii," *Voennaia mysl'* no. 10 (October 2020): 75.

High-Tech Futures 147

16. They assert that during research in the late 1980s, experts concluded it was important to integrate elements such as reconnaissance and electronic warfare, and develop command and control systems, i.e., "the same thing that is envisaged by the Western concept of 'network-centric warfare,'" but that these recommendations were not adopted. V. K. Kopytko and L. V. Kopylov, "O nekotorykh problemakh otechestvennoi voennoi nauki," *Voennaya mysl'* no. 9 (September 2013): 18.
17. Anatoly Kulikov, "Informatsiya kak strategicheskii resors" *Voenno-promyshlenniy kur'er* no. 14, 482, 10–16 (April 2013): https://vpk-news.ru/sites/default/files/pdf/VPK_14_482.pdf.
18. Kutishchev, "Voinyi budushchego: kakimi im biyt?," 63.
19. European states used the phrase "network-enabled capability."
20. See for example V. Adm. Arthur K. Cebrowski and John J. Gartska, "Network Centric Warfare: Its Origins and Future," *Proceedings* 124, no. 1 (January 1998); Adm. William A. Owens (USN), "The Emerging System of Systems," *Strategic Forum* no. 63 (February 1996).
21. According to N. N. Gerasimov and E. Shakirova, Western militaries consider the internet and information technologies to be the principal tools of NCW. They argue that NCW is actually "socionetwork-centric warfare." N. N. Gerasimov and E. Shakirova, "Sotsial'no-setetsentricheskiye voinyi sovremennosti: real'nost informatsionnoi epokhi," *Voennaya mysl'* no. 10 (October 2017): 79–87. Other analyses include A. E. Kondratyev, "Problemnye voprosy issledovaniya novykh setetsentrichskikh kontseptiivooruzhennykh sil vedushikh zarubezhnykh stran," *Voyennaya Mysl* no. 11 (November 2009): 1–74; A. E. Kondratyev, "Kogda 'setetsentrizm' pridet v rossiiskuyu armiyu?," *Vestnik Akademii Voennykh Nauk*, no. 2, 39 (2012): 120–125; V. Kutyishchev, "Voinyi budushchego: kakimi im byit?," *Armeiskii sbornik* no. 4 (April 2012): 60–63. See also V. Kovalyev, G. Malinetskii, and Y. Matviyenko, "Kontseptsiya 'setetsentricheskoi' voiny dlya armii Rossii: 'mnozhitel' sily' ili mental'naya lovushka?," *Vestnik Akademii Voennykh Nauk* no. 1, 50 (2015): 94–100; Saifetdinov, "Rol' voennoi nauki v sozdanii i razvitii avtomatizirovannoi sistemyi upravleniya Vooruzhennyimi Silami Rossiiskoi Federatsii"; A. V. Dolgopolov and S. A. Bogdanov, "Evolutsiya form i sposobov vedeniya vooruzhennoi bor'by v setetsentricheskikh usloviyakh", *Voennaya mysl'* no. 2 (February 2011): 49–58; A. Kopylov, "Budushchee za 'setetsentricheskimi voinami'?," *Armeiskii sbornik* no. 9 (September 2012): 53–58; A. Usikov, "Printsipyi i logika sovremennoi voinyi," *Armeiskii sbornik* no. 8 (August 2013): 48–50.

22. Aleksandr Khramchikin, "Pochemu SShA proigral voinu v Irake," *Nezavisimaya Gazeta* no. 10 (March 16, 2018): 6.

23. Vorobyev and Kiselev, "The role of military science in shaping the new look of the Russian Armed Forces," 43–44.

24. Nikolai N. Tyutyunnikov, *Voennaya mysl' v terminakh i opredeleniyakh v trekh tomakh. Tom 3: informatizatsiya vooruzhennyikh sil,* (Moscow:Pero, 2018), 159–160.

25. Bogdan I. Kazar'yan, "Operatsii, boevyie deistviya, setetsentrichenaya voina," *Voennaya mysl'* no. 2, (February 2010): 25–37.

26. Usikov, "Printsipyi i logika sovremennoi voinyi," 49.

27. Vorobyev and Kiselev, "The role of military science in shaping the new look of the Russian Armed Forces," *Voennaya mysl'* 2 (February 2011): 40–48.

28. Chekinov and Bogdanov, "O kharaktere i soderzhivanii voinyi novogo pokoleniya," 19.

29. G. Buturin and A. Evteev, "Voina i informatsionnyie tekhnologii," *Armeiskii sbornik* no. 5 (May 2013): 20.

30. Khamzatov, "Vliyanie kontseptsii setetsentricheskoi voiny na kharakter sovremennykh operatsii," 15-16.

31. A. V. Dolgopolov and Sergei. A. Bogdanov, "Evolutsiya form i sposobov vedeniya vooruzhennoi bor'by v setetsentricheskikh usloviyakh," *Voennaya mysl'* no. 2 (February 2011): 50.

32. Kazar'yan, "Operatsii, boevyie deistviya, setetsentrichenaya voina," 25–37.

33. Dolgopolov and Bogdanov, "Evolutsiya form i sposobov vedeniya vooruzhennoi bor'by v setetsentricheskikh usloviyakh," 56.

34. V. Kovalyev, G. Malinetskii, and Y. Matviyenko have expressed concern about China's development of "new forms and methods of warfare," including psychological and information operations, as well as command and control based on network-centric capabilities. V. Kovalyev, G. Malinetskii, and Y. Matviyenko, "Kontseptsiya 'setetsentricheskoi' voiny dlya armii Rossii: 'mnozhitel' sily' ili mental'naya lovushka?," 95.

35. V. Kovalyev, G. Malinetskii, and Y. Matviyenko, 96.

36. Dmitry Semyonov, "K upravleniyu oboronoi stranyi – pristupit'" *Krasnaya Zvezda,* December 2, 2014, http://archive.redstar.ru/index.php/newspaper/item/20252-k-upravleniyu-oboronoj-strany-pristupit.

37. Aleksandr Pinchuk, "V NTsUO sovershenstvuyut upravelenie oboronoi," *Krasnaya Zvezda,* September 30, 2019, http://redstar.ru/v-ntsuo-

sovershenstvuyut-upravlenie-oboronoj/; For further details about the Center, see Semyonov, December 2, 2014.

38. Aleksandr V. Khomutov, "Opit i perspektivyi ispol'zovaniya kontseptsii yedinoi informatsionno-kommunikatsionnoi seti v upravlenii voiskami," *Voennya mysl* no. 11 (November 2015): 21; Viktor Baranets, "My perelomili khrebet udarnym silam terrorizma," *Krasnaya Zvezda*, December 29, 2017, http://archive.redstar.ru/index.php/syria/item/35551-my-perelomili-khrebet-udarnym-silam-terrorizma

39. Pinchuk, "V NTsUO sovershenstvuyut upravelenie oboronoi."

40. Saifetdinov, "Rol' voennoi nauki v sozdanii i razvitii avtomatizirovannoi sistemyi upravleniya Vooruzhennyimi Silami Rossiiskoi Federatsii," 76–77.

41. For example, see V. V. Baranyuk, "Osnovnyie napravleniya sozdaniya edinogo informatsionnogo prostranstva VS RF," *Voennaya Mysl* no. 11 (November 2004): 29–34; V. K. Kopyitko and V. N. Sheptura, "Problemyi postroeniya edinogo informatsionnogo prostranstva Vooruzhennykh Sil Rossiiskoi Federatsii i vozmozhnyie puti ikh resheniya," *Voennaya Mysl* no. 10 (October 2011): 16–26; G. A. Lavrinov and A. A. Chumichkin, "Opyt sozdaniya edinogo informatsionnogo prostranstva dlya resheniya zadach tekhnicheskogo osnashcheniya Vooruzhennykh Sil Rossiiskoi Federatsii," *Vestnik Akademii voennykh nauk* 1, no. 26 (March 2009): 76–81.

42. Viktor Khudoleev, "Voennaya nauka smotrit v budushchee," *Krasnaya Zvezda*, March 26, 2018, http://redstar.ru/voennaya-nauka-smotrit-v-budushhee/.

43. Anatoly Ya. Chernysh and Vladimir V. Popov, "Ob evolyutsii teorii I praktiki edinogo informatsionnogo prostranstva I pervoochered-nykh merakh po ego razvitiyu v interesakh povyisheniya effektivnosti upravleniya natsional'noi oboronoi Rossiiskoi Federatsii," *Voennaya Mysl* no. 9 (September 2019): 47–54.

44. N. I. Sidnyaev, "Setetsentricheskiye upravlyayushchie sistemy i boevyie operatsii," *Voennaya Mysl* no. 12 (December 2021): 63 –64.

45. Vladimir I. Slipchenko, "Novaya forma borbyi. V nastupivshem veke rol' informatsii v beskontaktnykh voinakh budet lish vozrastat," *Armeiskii sbornik* no. 12 (December 2002): 30–32.

46. Yury E. Donskov, S. N. Zhikharev, and A. S. Korobeinikov, "Primeneniye sil i sredstv radioelektronnoi borbyi pri zashchite nazemnykh obektov ot sredstv vozdushno-kosmicheskogo napadeniya," Voennaya mysl' no. 8 (August 2015): 24–29.

47. Yury E. Donskov, V. I. Zimarin, and B. V. Illarionov, "Podkhod k postroeniyu system radioelektronnoi borbyi v usloviyakh realizatsii setetsentricheskikh kontseptsii razvitiya vooruzhennykh sil," *Voennaya Mysl* no. 2 (February 2015): 48.
48. Yury E. Donskov, A. S. Korobeinikov, and O. G. Nikitin, "K voprosu o prednaznachenii, meste i roli voisk radioelektronnoi borbyi v armeiskikh operatsiyakh," *Voennaya Mysl* no. 12 (December 2015): 20 –24.
49. Aleksandr Tikhonov, "Sferyi sosredotocheniya osnovnykh usilii," *Krasnaya Zvezda*, March 1, 2021, http://redstar.ru/sfery-sosredotocheniya-osnovnyh-usilij/. See also Vladimir Pylaev, "Ikh zadacha protivodeistvovat sredstvam napadeniya," *Krasnaya Zvezda*, April 19, 2021, http://redstar.ru/ih-zadacha-protivodejstvovat-sredstvam-napadeniya/.
50. Vladimir A. Vinogradov, "Kharakternyie chertyi sovremennykh obshchevoisovykh operatsii," *Voennaya mysl'* no. 1 (2001): 23–26.
51. Ivan Vorobyev and Valery A. Kiselyev, "Vyisokotochnoye srazheniye," *Voennaya mysl'* no. 11 (November 2006): 15–22.
52. Chekinov and Bogdanov, "Evolyutsiyasushchnosti i soderzhaniya ponyatiya voina v XXI stoletii," 36–37.
53. See, for example, Kruglov and Shubin, "O vozrastayushchem znachenii uprezhdeniya protivnika v deistviyakh," 27–34.
54. Vladimir B. Zarudnitsky, "Faktoryi dostizheniya pobedyi v voennykh konfliktakh budushchego," *Voennaya mysl'* no. 8 (August 2021): 34–47.
55. They also argue that the adoption of hypersonics means that future war will be global in nature. Stepshin and Anikonov, "Razvitiye vooruzheniya, voennoi i spetsial'noi tekhniki i ikh vliyaniye na kharakter budushchikh voin."
56. Vorobyev and Kiselyev, "Vyisokotochnoye srazheniye."
57. Oleg V. Tikhanychev, "O roli sistematicheskogo ognevogo vozdei'stviia v sovremennykh operatsiiakh," *Voyennaya Mysl'* no. 11 (November 2016): 16–20.
58. Ivan Vorobyev, "Informatsionno-udarnaya operatsiya," *Voennaya mysl'* no. 6 (June 2007): 14–21.
59. See for example Tikhanychev, "O roli sistematicheskogo ognevogo vozdei'stviia v sovremennykh operatsiiakh," 16–20.
60. President of Russia (website), "Poslaniye Prezidenta Federal'nomu Sobraniyu," February 20, 2019, http://kremlin.ru/events/president/news/59863.
61. Ibid..

62. Oleg Falichev, "Vosem makhov ne predel," *Voenno-promyshlennii kurer,* February 8, 2021, https://vpk-news.ru/articles/60778.
63. Trine Jonassen, "Russia test fires submarine-launched hypersonic Tsirkon missile for the first time," *High North News,* October 5, 2021, https://www.highnorthnews.com/en/russia-test-fires-submarine-launched-hypersonic-tsirkon-missile-first-time.
64. President of Russia (website), "Meeting with Defence Ministry leadership and heads of defence industry enterprises," November 2, 2021, http://en.kremlin.ru/events/president/transcripts/67056.
65. Viktor Baranets, "My perelomili khrebet udarnym silam terrorizma," *Krasnaya Zvezda,* December 29, 2017, http://archive.redstar.ru/index.php/syria/item/35551-my-perelomili-khrebet-udarnym-silam-terrorizma.
66. Khudoleev, "Voennaya nauka smotrit v budushchee."
67. Oleg Pochinyuk,"S uchyotom siriiskogo opyita."
68. See for example the detailed analysis of US use of UAVs in Iraq, Afghanistan, and elsewhere in Nikolai Novichkov, "Zona deistvii – ot Afghanistana do Afriki," *Voenno-promyshlennii kurer,* February 13, 2012, https://vpk-news.ru/articles/8619.
69. A. V. Shubin, I. V. Kot, and A. A. Kuzenkin, "Izmeneniye kontseptualnyikh podkhodov k primeneniyu aviatsii v voinakh budushchego na primere karabakhskogo konflikt," *Voennaya mysl'* no. 9 (September 2021): 46.
70. Aleksei Ramm, "Kuda letit bespilotnaya aviatsiya," *Nezavisimoye Voennoye Obozrenie,* January 21, 2021, https://nvo.ng.ru/armament/2021-01-21/1_1125_aviation.html?mc_cid=be9ee0c7b1&mc_eid=b7db4de0aa.
71. V. V. Repin, "Razvitiye teorii primeneniya voisk protivovozdushnoi oborony Sukhoputnykh voisk," *Voennaya mysl'* no. 11 (November 2018): 26–29.
72. A number of analyses of military technology and AI examine what the US has been doing. For example, Vladimir Ivanov, "Iskusstvenyi intellekt – osnova budushchykh srazhenii," *Nezavisimoe voennoe obozrenie* no. 42 (November 12, 2019): https://nvo.ng.ru/nvo/2019-11-15/1_1070_al.html; Vasily M. Burenok, "Iskusstvennyi intellect v voennom protivostoyanii budushchego," *Voennaya Mysl* no. 4 (April 2021): 106–112; D. V. Galkin, P. A. Kolyandra, and A. V. Stepanov, "Sostoyaniye i perspektivi ispolzovaniya iskusstvennogo intellekta v voennom dele," *Voennaya Mysl* no. 1 (January 2021): 113–124; V. K. Istanov, "Nazemnaya robototekhnika Soyedinenniykh Shtatov Ameriki, Germanii, Kitai:

sostoyaniye i perspektivyi razvitiya," *Voennaia mysl* no. 1, (January 2022): 143–156.

73. Andrei Kolesnikov, "Zasluzhennyie deyateli iskusstvennogo," *Kommersant,* Novermber 10, 2019, https://www.kommersant.ru/doc/4 154715.

74. Quoted in Andrei Ilnitsky and Alexander Losev, "Iskusstvennyi intellekt – ugrozy i vozmozhnosti," *Arsenal Otechestva* 42, no. 4 (2019): 4–7.

75. President of Russia (website), "Artificial Intelligence conference," November 12, 2021, http://en.kremlin.ru/events/president/transcripts/ 67099.

76. D. V. Galkin, P. A. Kolyandra, and A. V. Stepanov, "Sostoyaniye i per-spektivi ispolzovaniya iskusstvennogo intellekta v voennom dele."

77. Zarudnitsky, "Kharakter i soderzhaniye voennyikh konfliktov v sovre-mennykh usloviyakh i obozrimoi perspective," 43.

78. Burenok is president of the Russian Academy of Rocket and Artillery Sciences and from 2002 worked at the 46th Central Research Institute, focusing on the development of weapons and military equipment. He has published widely in the military theoretical literature journals, both in *Voennaya Mysl, Voenno-promyshlennii kurer,* and more specialist publications. See for example, Vasily Burenyok, "Arsenal shestogo pokoleniya," *Voenno-promyshlennii kurer,* February 26, 2019, https://vpk-news.ru/articles/48628.

79. Vasily M. Burenok, "Iskusstvennyi intellect v voennom protivostoyanii budushchego," *Voennaya Mysl* no. 4, (April 2021): 112.

80. See https://fpi.gov.ru/.

81. "Tsentr robototekhniki MO RF," *Voenno-promyshlennii kurer,* August 12, 2013, https://vpk-news.ru/articles/17061; see also Yuri Avdeev, "Robototekhnik mnozhit boevoi potentsial" *Krasnaya Zvezda,* July 4, 2018, http://redstar.ru/robototehnika-mnozhit-boevoj-potentsial/.

82. For further details, see Military Technopolis Era, https://www.era-tehnopolis.ru/.

83. "Minoboronyi k 1 dekabrya sozdast upravlenie dlya raboty s iskusstvennyim intellektom," *Tass,* May 31, 2021, https://tass.ru/armiya-i-opk/11515073.

84. The Uran-6 is designed for mine clearance and can do the work of an entire platoon. The Uran-9 and Uran-14 can be used in combat. Dmitry Litovkin, "Robotyi idut na voinu," *Nezavisimoye Voennoye Obozrenie,* April 15, 2021, https://nvo.ng.ru/nvoweek/2021-04-15/1_1137_robots.html.

85. "Na ucheniyakh 'Zapad-2021' vpervyie primenili robotyi 'Uran-9' i 'Nerekhta'," *Tass*, September 13, 2021, https://tass.ru/armiya-i-opk/12 372947.

86. Litovkin "Robotyi idut na voinu."

CHAPTER 5

UNDERMINING THE
WILL TO RESIST

A distinctive focus of Russian military thought in recent decades has
been on nonmilitary means of achieving strategic objectives, particularly
by manipulating a state's population and undermining it from within,
weakening its will to resist. New technologies and information are
perceived to be both critical enablers and key instruments of these
nonmilitary approaches, contributing to a significant shift in the character
of conflict. This chapter explores Russian views on nonmilitary means
of meeting strategic objectives and how Russia perceives other actors (in
particular the US and its allies) seek to achieve their goals. In particular,
it examines Russian concern about (and interest in) nonmilitary means
of destabilization, such as "controlled chaos" and the perceived threat
from color revolutions, which are considered to be elements of a strategy
deliberately employed by the West to undermine its strategic competitors.
The views of soft power and "controlled chaos" as distinct features
of contemporary and future wars are clearly expressed in the Russian
military theoretical debate and are central to understanding the Russian
view of contemporary conflict. Internal destabilization and the threat

from "color revolutions" have become a recurrent leitmotif in Russian strategic thought and writing on the character of future conflict. This reflects a view across the Kremlin and beyond that events such as the Arab Spring uprisings in 2011, demonstrate the threat from anti-government demonstrations, which can quickly turn into a state of chaos —or "controlled chaos"—and threaten a state. Information warfare (which includes cyber operations) forms a key part of this, constituting a critical enabler and important means, and will be investigated in detail, along with psychological operations. These means—tools of indirect action— seek to target society, along with the decision-making of a state, thereby undermining its ability to resist an opponent. These means seek to create uncertainty and unpredictability, exploiting existing divisions.

A number of Russian experts have argued that future warfare will be conducted increasingly by nonmilitary means and indirect action, including internal destabilization, subversion, and information confrontation. For example, Gorbunov and Bogdanov have set out new forms of conflict in which military means are either not used or have no predetermined role, arguing that the weakening of a state in order to deprive it of the will to resist and/or political destruction is a key objective of contemporary aggression. In their view, this could be achieved by undermining the state internally through a variety of possible means including inciting opposition groups to take action and provoking national differences and inter-ethnic conflicts.[1] Andrei V. Kartapolov identifies the essence of so-called "indirect actions" (echoing Liddell Hart's indirect approach), which he differentiates from classic forms of warfare, as the manipulation of internal contradictions within a state in pursuit of strategic objectives.[2] Kurochko differentiates between "classical" and "non-classical" war, asserting that in a classical war, the target is an adversary's material resources and their physical defeat. By contrast, non-classical wars target the consciousness of the individual and society with the use of non-physical violence in an irregular indirect manner, using informational-psychological and economic means, among others.[3]

INTERNAL DESTABILIZATION AND CONTROLLED CHAOS

Russia's national security is perceived to be under threat from internal instability, subversion, and regime change initiated by external actors, targeting society's will to resist rather than state structures. A country's population and social institutions are considered to be more susceptible to subversive manipulation than official state structures, as well as being critical to a state's ability to withstand an armed attack. According to this view, the principal objective of a hostile actor is no longer the destruction of an opponent's military and the physical seizure of territory; rather, the aim is to gain the ability to influence an opponent's way of life, values, behavior, and, ultimately, decision-making, weakening the ability of the actor to resist hostile action. This reflects the principal aims identified in Slipchenko's concept of sixth-generation warfare, which are change in political rule and the obliteration of an adversary's economic potential rather than the seizure and holding of territory. Advanced information and communications technologies are deemed to play a central role in nonmilitary approaches, facilitating an actor's ability to covertly influence a target group, such as a specific section of a population, an organization, or individuals.[4] Computers and mobile phones have become tools of information warfare, allowing actors to spread disinformation and incite fear in order to sow division and influence a population, undermining trust in political elites and state institutions. Troll factories and bots facilitate anonymity, and social media platforms, designed to enable information-sharing across wide, disparate networks, provide an ideal medium to disseminate challenging views.

The discussion about destabilization and controlled chaos is not new; aspects of it featured in Russian military thinking long before the twenty-first century—for example, Evgenii Messner's work on wars of rebellion (*miatezhevoiny*).[5] This stems partly from the fact that the formative experience of the Soviet state, which was a product of revolution and civil war, prompted persistent unease about counterrevolution, internal subversion, and any perceived challenge to regime stability. Enduring

Soviet concerns about the potential for counterrevolution and Western attempts to undermine the USSR using indirect, nonmilitary means were evident during the 1968 Prague Spring, when the Warsaw Pact broadened its interpretation of security to incorporate efforts to subvert socialism from within and to protect the state against external military attack.[6]

This enduring concern about the possibility of internal subversion and interference in Russia's domestic affairs remains highly visible in ongoing theoretical discussions about the character of conflict and how states achieve their strategic objectives in the twenty-first century. Aleksandr Bartosh, a member of the Academy of Military Sciences and former military adviser to the Foreign Ministry, has written extensively on nonmilitary means of influence, arguing that subversion is the principal dimension of Western hybrid warfare. He has suggested that subversion is utilized to undermine an adversary's ability to resist through the use of tools such as economic and information-psychological operations, terrorism, nationalism, militant groups and special operations forces.[7] According to this view, interstate confrontation and conflict have become all-encompassing, centered around nonmilitary means. Bartosh suggests that education of the population is vital in order to be able to "seize the initiative...in information and psychological confrontation." He concludes that the West's recognition of the extent of Russian military power has forced it to seek "workarounds" to undermine the state and that the country's armed forces should be capable of tackling a wide range of both military and nonmilitary threats, including political, economic, and informational tools, implemented by "the widespread use of the protest potential of the population and special operations forces."[8]

The focus on undermining an adversary's will to resist echoes both Sun Tzu, who stated that supreme excellence in war consists of breaking an adversary's resistance without fighting, and Liddell Hart's indirect approach, which sought to reduce resistance from an adversary, particularly its decision makers, by taking the "line of least resistance" in physical terms and the "line of least expectation" in psychological terms.[9] In the contemporary era, Russian theorists consider the information space to be critical, particularly the informational-psychological realm, which is considered to be directly connected to morale and an adversary's ability to resist—information confrontation enables an actor to "neutralize" an opponent without using weapons or resorting to overt military means. An indirect approach is used to obtain opportunities to ensure the development of the state and contribute to the "weakening and elimination of military dangers and threats."

An article by Vorobyev and Kiselyev in 2006 examined the "strategy of indirect actions," exploring the historical roots of such an approach, as well as its evolution in the contemporary era. They express a strong conviction that a strategy of indirect actions, which had previously been sidelined by a "strategy of force," was becoming indispensable for military leaders in the twenty-first century. Chekinov and Bogdanov returned to the concept of indirect action in 2011, arguing that the increasing use of nonmilitary means to achieve strategic objectives in the contemporary security environment had been driven by the "catastrophic consequences" of conflict involving modern weaponry. According to them, nonmilitary means are related to the concept of indirect actions: both are focused on the defeat of an adversary by indirect, often ambiguous, methods.[10] Kartapolov notes the key role of information in indirect actions, arguing that "information confrontation comes to the fore" in such actions, facilitating the destabilization of a state and influencing the decision-making process in order to unseat the existing regime.[11] In his view, information operations can have a comparable effect on mass consciousness to a large-scale military deployment, with falsification or the distortion of information—that is, disinformation—being the most effective means of information confrontation:

> Neither the leadership nor the population of the state being targeted by information 'pressure' is initially aware of what is happening. The resulting confrontation…is not perceived by the masses as war, since there are no obvious signs of external aggression. The front between the warring parties takes place, first of all, in the

public consciousness and in the head of every person. In these 'new wars', there are not the mass casualties as witnessed during WW1 and WW2; nevertheless, aggression is directed predominantly towards civilian populations.[12]

Western states are perceived to have been employing an indirect approach and nonmilitary means in pursuit of their strategic objectives. A number of Russian analyses[13] refer to work by a US analyst, Steven R. Mann, who published a paper entitled "Chaos Theory and Strategic Thought" in 1992. Mann's work analyzes the impact of nonlinear paradigms such as chaos theory and criticality on strategic thinking and national security, which were popular topics of scholarly discussion in the early post–Cold War era.[14] He labeled this the theory of self-organized criticality, identifying it as the latest trend in the development of the theory of dynamic nonlinear systems. Bartosh traces the concept of "controlled chaos" to the US and references Mann in particular, describing him as one of the developers of the concept and suggesting that information and psychological expertise is used to undermine strategic competitors and adversaries:

> The United States considers...chaos 'manageable' and sees in it a new instrument for promoting its national interests under the pretext of democratising the modern world and spreading liberal values. Other countries, including Russia, view the use of 'controlled chaos' technologies as a...disaster that can lead to global catastrophe.[15]

Mann contended that ideology is a "virus," which can be used to shape an adversary's population and a favorable international order by spreading chaos, and called for an increase in government support for organizations such as the National Endowment for Democracy and the US Information Agency, as well as educational and exchange programs:

> The US should move to the ultimate biological warfare and decide, as its basic national security strategy, to infect target populations with the ideologies of democratic pluralism and respect for individual human rights. With a strong American commitment,

enhanced by advances in communications and increasing ease of global travel, the virus will be self-replicating and will spread in nicely chaotic ways.[16]

Although there is no evidence of Mann's recommendations being adopted as policy or of any direct link between his line of thinking and specific US government action, there is little doubt that the US actively sought to pursue the spread of values such as liberal democracy. During the 1990s the centerpiece of American foreign policy was the claim that promoting the spread of democracy would also promote global peace and security. Bill Clinton's 1994 National Security Strategy was the first explicit post–Cold War articulation of democracy promotion, a theme echoed in George W. Bush's 2002 document, which was premised on the need to promote democracy in all countries and "a distinctly American internationalism that reflects the union of our values and our national interest," emphasizing nonmilitary forces of power.[17] Thus, the promotion of democracy was a central plank of US foreign policy articulated in the 2002 National Security Strategy. This fueled Russian fears about the possibility of foreign interference in their internal affairs: the promotion of democracy came to be viewed as part of a new US-led approach to warfare, which involved the internal destabilization of rival states through largely nonmilitary means such as democratization and regime change to achieve fundamental security objectives. "Color revolutions" are one supposed form of internal destabilization that has received a lot of attention from Russian theoreticians. The term emerged in reference to a series of anti-government protests and popular uprisings that occurred across the post-Soviet space in the early to mid-2000s, intended to result in nonviolent regime change. A narrow definition only includes the 2003 Rose Revolution in Georgia, 2004 Orange Revolution in Ukraine, and 2005 Tulip Revolution in Kyrgyzstan. However, there is a level of ambiguity around the Russian interpretation of the term, and it has been applied to a wide range of events from the 2000 "Bulldozer" revolution in Serbia to the Arab Spring in North Africa and the Middle East in late 2010 and early 2011.

The threat posed to Russia from "color revolutions" is an enduring theme in Russian security discourse, and they are characterized as an externally sponsored form of regime change and a challenge to the Westphalian model of sovereignty, part of a Western conspiracy to usurp Russian influence. This was apparent in a speech to the Russian Duma in May 2005 by then FSB director Nikolai Patrushev, who accused "certain political forces" from the West of acting in the "worst traditions" of the Cold War and claimed that foreign intelligence services were using "non-traditional methods" to covertly manipulate states from within, using nongovernmental organizations as a cover for a variety of influence operations.[18] A common theme in Russian writing on the color revolutions phenomenon is the notion that they are a manifestation of the struggle for dominance of the post-Soviet space that has been taking place between Russia and the United States, and have been deliberately engineered to isolate Russia from its neighbors. They are equated with the implementation of a coup d'état, part of a Western strategy to preserve power and influence through the "creation and manipulation of conditions of chaos" around the world in order to undermine competitors and destabilize them from within.[19] The United States is perceived to be the mastermind of this strategy, which it pursues around the world in the form of direct and covert military intervention as well as in the form of color revolutions, which are part of a wider strategy of democratization and, consequently, regime change. Gene Sharp's *From Dictatorship to Democracy*[20] is pinpointed as an important instrument of US-sponsored regime change and the inspiration for a number of nonviolent uprisings, including the Arab Spring. It is referred to in a number of Russian analyses of the color revolution phenomenon, including G. Filimonov, N. Danyuk, and M. Yurakov, who describe Sharp as the architect of the theory and practice of color revolutions.

Gareev was one of the principal exponents of the potential threat to Russian national interests from externally-sponsored internal instability and deliberate destabilization. For over a decade he argued that large-scale war was unlikely, because other, nonmilitary, methods of achieving political goals have been developed, such as economic sanctions, diplomatic pressure, and information war. Speaking in 2004, he asserted that it is possible to conquer one country after another through subversive actions from within, referencing events in Iraq to support his argument.[21] Gareev warned that passivity in the fields of politics, diplomacy, and information had been Russia's greatest weakness, arguing that if threats and defense problems were not effectively dealt with by political means, then adversaries would take advantage of them. According to this thesis, popular uprisings, civil unrest, and controlled chaos are all specifically designed by an adversary to undermine the stability of a state and collapse it from within, and represent a key threat to Russia:

> On the whole we see an erosion of the lines between military and nonmilitary means of international confrontation, as well as an increase in the possibility of covert, asymmetric, violent military and nonmilitary influence.[22]

The tone of Gareev's article needs to be understood within the context of both the domestic and international instability that had occurred during 2011: There were protests in Russia during the run-up to the Duma elections of December 2011, which followed the Arab Uprisings of 2011 and the post-Soviet color revolutions of 2003 to 2005. The 2011 CGS, Nikolai Makarov, voiced apprehension about the potential for internal instability in Russia instigated by external interference, warning that certain states were seeking to use the "technology" of color revolutions to achieve their own strategic objectives. This echoes a prevalent belief amongst security elites that the instability and conflict in the Middle East and North Africa demonstrated the unquestionable threat posed by anti-government demonstrations, which could quickly turn into a state of chaos, "controlled chaos," and pose an existential threat to the security of a state. In 2012, Putin set out his conviction that the West used various methods—political and economic—to destabilize and undermine Russia's neighbors and therefore ultimately the Russian Federation:

Before our eyes, new regional and local wars are breaking out. Zones of instability and artificially incited, controlled chaos. And there are targeted attempts to provoke such conflicts in the immediate vicinity of the borders of Russia and our allies.[23]

The prominent statements and writing on the perceived threat from color revolutions from CGS General Valery V. Gerasimov in 2013 are the most obvious manifestation of this. His now infamous article in *VPK*, "The Value of Science in Foresight," summarized a speech he had made to the annual meeting of the AVN in January of that year in which he had reflected upon the changing character of conflict, implicitly encouraging members of the AVN to address these challenges in their future work. Gerasimov built on the themes that had been developed by other writers and analysts, such as those outlined above, asserting that the experience witnessed in the "so-called color revolutions in North Africa and the Middle East confirm that a prosperous state can in a matter of months...become a victim of foreign intervention and sink into an abyss of chaos, humanitarian disaster and civil war." Using the events of 2011 in North Africa and the Middle East as an example of how twenty-first-century conflict is changing, Gerasimov noted that there has been a shift toward the "widespread use of political, economic, informational, humanitarian, and other nonmilitary measures implemented along with the manipulation of the protest potential of the population. All this is supported by hidden military measures, including the implementation of information operations and special forces..."[24] This echoes Gareev's 2012 article, discussed earlier, in which he warns about the erosion of the lines between military and nonmilitary means in international confrontation. Gerasimov was deeply critical of the United States and NATO for destabilizing the international environment with these new, predominantly nonmilitary, methods to advance their own national and coalition interests "under the pretext of spreading democratic values." Color revolutions were perceived to be the principal instrument of this new approach:

> [Color revolutions] are based on political technologies involving external manipulation of the protest potential of the population in combination with political, economic, humanitarian and other nonmilitary measures. Over the last decade, a wave of such 'color' revolutions has been initiated by the United States in the post-Soviet space, in North Africa and in the Middle East.[25]

A. N. Belsky and O. V. Klimenko appear to invent a new discipline, "color revolution engineering," claiming that algorithms are used to foment unrest and destabilize a country. In an article published in 2014, they outlined ways to counter color revolutions, which they characterized as a "period of street riots and popular unrest sponsored by foreign non-governmental organizations and commonly succeeding in the overthrow of the ruling political regime."[26] In their opinion, a key objective of the color revolutions that took place in the post-Soviet space was to "split the countries within the post-Soviet space from each other and hem in Russia with neighbors that were far from friendly." Thus, they concluded that Russia itself was a target for forces seeking to isolate it. In 2019 Gerasimov set out how Russia intended to respond to what he described as the Pentagon's new "Trojan-horse" strategy, which combined active use of fifth columnists and manipulation of a state's "protest potential" in order to destabilize the country internally, with long-range precision strike against critical targets. The Russian CGS outlined a strategy of active defense that had been developed by military scientists working in conjunction with the General Staff with the aim of neutralizing aggression by potential adversaries. According to Gerasimov, this strategy enabled Russia to "preemptively neutralize" perceived threats to the state in line with the country's defensive posture set out in the Military Doctrine.[27]

The Russian understanding of "color revolutions" and the perceived threat to national security stemming from such events reflects a broader assessment of the character of conflict in the twenty-first century and how actors pursue their strategic objectives. As discussed earlier, a number of experts perceive a shift away from physical destruction and the seizure of enemy territory toward the "psychological crushing" and exhaustion of an adversary, with the intention of weakening their ability to resist.[28] This echoes Andrei Snesarev, who advocated "killing the spirit of the individual soldier," thereby depriving militaries of the will to fight, rather than killing soldiers themselves.[29] V. Potekhin and Yu Gromyko have developed the idea of a "conscient" war, one which is psychological in form and seeks to change (or destroy) the values held by a particular society. According to their interpretation, a "conscient" war implies that the world has entered a new phase of struggle, a competition between different cognitions and understanding, in which certain approaches "simply have to be destroyed."[30] Although this type of conflict, between different value systems, is not new, the idea of the principal arena for conflict and confrontation between actors shifting to the cognitive realm, rather than the physical, means that the cognitive realm has become a battlespace, where victory is won by the domination of ideas and narratives, rather than physical territory. The use and manipulation of information plays a key role in this and has been a central focus of Russian military and strategic thought for many years.

CONFRONTATION IN THE INFORMATION REALM

In 2018 Russian Deputy Defense Minister Yury Borisov argued that in the contemporary era, "battles are not played out on the battlefields, they are first played out in the information space. Whoever is able to control it, who is able to organize opposition in the right way, becomes the winner today."[31] The information domain has been a central focus of both the Russians and the Soviets. A number of Russian analysts have noted that modern society cannot function without information, which can be used for both creative and destructive purposes, and echo Ogarkov's statement from the 1980s that information would comprise the battlefield of the future. This view is reiterated in Gerasimov's 2013 *VPK* article, which states that the role of nonmilitary means (including informational means) had increased and in some cases "significantly surpassed the power of weapons in their effectiveness."[32] Similarly, Vladimir B. Zarudnitsky has argued that information confrontation (*informatsionnoye protivoborstvo,* the commonly used Russian term for information warfare) constitutes one of the most effective nonmilitary means of achieving strategic objectives. He refers to an arsenal of "uncertainty strategies" that utilize information, including propaganda, sabotage, disinformation, and political destabilization, all of which have become increasingly effective in the pursuit of an actor's strategic goals.[33] Changes in information technology and communications, as well as the growing interconnectedness of societies, has accelerated the importance of information within the character of conflict. Information warfare is not new, but the manipulation of interconnected, information-rich environments means that it is becoming increasingly difficult to distinguish between friend and foe, and the traditional binary distinction between war and peace is often unclear. Information can be used in a wide variety of ways to influence an adversary: it can be used to divide populations, undermine a state's will to resist, confuse an adversary, and increase support for Russia and its leadership, among others. Information warfare forms a key part of this.

Russian sources tend to use a variety of different terms when referring to "information warfare" and there has been a wide-ranging debate about terminology. One of the most common terms is "information confrontation," which appears to correlate with the Western understanding of information war. Other terms used include information battle and information war, reflecting the perceived difference between *bor'ba* and *voina* discussed in chapter 1. Russian thinkers categorize information confrontation as either technical and psychological. When conducting the first, the main objects of influence and protection are information and technical systems (communication systems, telecommunication systems, radio electronic means, etc.). As discussed earlier (vis-à-vis NCW), Russia has made significant improvements to its C4ISR ability, improving its own situational awareness, as well as its ability to challenge the awareness of adversaries (for example, through EW). This links to the desire to seize and maintain the initiative during the IPW and the use of surprise. The military focus on information superiority reflects the growing conviction among Russian political and military elites that information confrontation is one of the fundamental ways in which states (and other actors) compete in the twenty-first century. During information-psychological confrontation, the psyche of the political elite and the population of the opposing sides is influenced, which in turn impacts the system for the formation of public consciousness, opinion, and decision-making.[34] Thus, notwithstanding the debate over terminology and the broad range of activities covered by "information confrontation," the objective of all of these is similar: to deprive an adversary of the ability to fight, exploiting existing vulnerabilities and divisions within society. Colonel Anatoliy Nogovitsin, former Deputy Chief of the General Staff, defines information warfare as:

> Confrontation between countries carried out in the information sphere in order to damage information systems, processes, means, critical infrastructure and to weaken the political and social systems, as well as the psychological manipulation of military personnel and civilians on a mass scale in order to completely destabilise the enemy's society and country. The main goal of information warfare is to destroy the foundations of the national identity and the way of living of the hostile country.[35]

Thus, he clearly differentiates between information-technical and information-psychological confrontation, with the latter being aimed at undermining a state's cognitive abilities by spreading chaos. He identifies a number of characteristics of information confrontation that, in his view, distinguish it from other forms of warfare, including the low costs associated with the development and use of information warfare, an increased role for managing perceptions, and the vulnerability of a state as a result of growing reliance on IT. He warns that in contemporary warfare victory is ensured by the preemptive acquisition of information superiority, which must occur before military superiority can be achieved, because "defeating the enemy is essentially a psychological act." This echoes a celebrated assertion by Suvorov that surprising an adversary is a partial victory: penetrating an adversary's mind is the principal and most difficult task for a military commander. Keir Giles notes that the Russian concept of information warfare covers a vast range of activities and processes, as well as a wide array of methods, including intelligence, counterintelligence, military deception (*maskirovka*), disinformation, psychological operations, propaganda, cyber operations, and electronic warfare.[36] Furthermore, Russia does not distinguish cyber warfare from information warfare; they are considered to be part of the same broad spectrum: the term "information space," rather than cyber space, is used in Russian.

The Russian focus on the role of information and technology has partly grown out of a recognition that although the US remains technologically superior, a key vulnerability of liberal democracies is their core values, including freedom of expression and speech. There is also a common belief that Russia is already involved in "information confrontation" with Western states:

> Modern Russia, which occupies one-eighth of the world's land mass and has one of the most powerful armies in the world, is involved in a number of information conflicts with various countries. Our main opponents are the so-called 'Western countries', primarily the United States, as well as their NATO allies.[37]

Information warfare, seen as being implemented by other states, is considered by some to be the principal threat to society as a whole and the country's security services. Russia is considered to be under pressure from information warfare as other states seek to project a negative image of the country around the world as part of efforts to shift the correlation of forces.[38] Thus, information warfare is perceived to be a powerful means of indirect, nonmilitary influence over open societies, partly because it is very difficult to attribute. Some Russian observers discern growing confrontation in the global information space, driven by the US and its allies, who are seen to be "manipulating the public consciousness" of their own domestic populations, as well as those of their adversaries.

This is considered to reflect a change in contemporary conflicts from a classical linear paradigm to a non-linear, new type of war.[39] Aleksandr N. Limno and M. F. Krysanov suggest that the growing role of information warfare in contemporary military operations, particularly information confrontation, as a result of the rapid "informatization" (i.e., reliant on information technologies) of all spheres of human activity, could trigger a profound change in the theory and practice of military art.[40] A number of military theorists support this, arguing that victory is only possible through the superior control of information and intellectual potential compared one's adversaries.[41]

A lot of attention has been focused on the approach of both China and the US to information warfare. In 1999, E. N. Dezhin analyzed the Chinese view of information warfare, which perceived a future war being triggered by an error in global IT networks: those involved would not be soldiers, and information systems would play a key role, impacting on all aspects of life, including the military, politics, and the economy.[42] From his analysis, Dezhin surmised:

> The age of information has brought about fundamentally new means, methods, and conceptions of warfare. It may affect the material world around us little or not at all yet leave behind completely ruined information networks. The enemy can be defeated or weakened through blackmail and other active measures effected through information. There will be no bloodshed but a bloodless confrontation of information systems.[43]

Migunov echoed this view in his 2008 analysis of China's approach to information warfare, which is defined as a "transition from a mechanized war of an industrial society to a war of decisions and management style, a war of knowledge and intelligence," in which the target is psychology. Migunov noted that, in China's view, information war is the "war of the future."[44] Like Russia, China divides information warfare into two types: information-technical and information-psychological.[45] Thus, Russian theorists perceive Chinese views of information warfare as being very similar to their own.

During the late 1990s and early 2000s, there was a significant debate within Russian expert circles about "information warfare" and its utility both as a theoretical concept and as a tool of statecraft. Many analysts noted that the use of information to undermine an adversary is not a new phenomenon, pointing to the Cold War as a prime example.[46] Oleg N. Kalinovsky warned against overemphasizing the impact that information could have, asking whether the information campaign against Milosevic in 1999 would have worked without the concurrent bombing campaign and economic blockade.[47] Sergei A. Komov wrote a number of articles on information warfare in the late 1990s, focusing on the role of information in contemporary warfare and the impact of digitization. Although he emphasized that information had been an object of warfare throughout history, he warned that the growing reliance of armed forces on information technologies was creating a fundamentally new military situation, pointing to the development of electronic warfare and the use of IT in weapons systems.[48] He divided the methods of information warfare into three major categories: kinetic/use of force, cognitive, and combined. Cognitive methods included deception and decoy, division, exhaustion, appeasement, intimidation, provocation, and reflexive control; it is examined in more detail in the following paragraphs.

Komov's work was developed by Rodionov, who stated that a key task of military science was to understand the forms and methods of information warfare in order to facilitate the effective use of forces and assets.[49] Attaining information superiority over an adversary was identified as the principal objective of information warfare, which, according to Rodionov, has a number of forms: information warfare operations, information battles, information actions, and information strikes. These different forms map across the strategic, operational and tactical levels, with IW operations constituting an aggregate of information struggles, actions, and strikes, in pursuit of information superiority over an adversary either in-theater or at the strategic level. Information actions include specifically targeted impacts, manipulation of information, information blockades, and defensive protection measures. An information-warfare strike is characterized as a coordinated attack on critical elements of an adversary's command and control system in pursuit of information superiority.[50]

Orlyansky echoed Komov in several detailed articles on information warfare published in the early 2000s, arguing that it had been waged for centuries and that "while the essence and goals of this struggle remain unchanged, the means, content, forms and methods of its conduct... have been improved."[51] This emphasizes the impact of new technologies, such as the internet. Nevertheless, he subsequently warned against overemphasizing the potency of information as an instrument of war and influence, echoing the views of people like Kalinovsky. Writing in 2008, Orlyansky noted that information has always played a supporting role in war and conflict: "for all the importance of information, it does not replace and, possibly, will never replace weapons, it will not become the main means of fighting to defeat...Information has not yet become a means capable of influencing a person as effectively as modern weapons."[52] He cited work by the Russian Academy of Military Science, which states that "waging a victorious information war is still an intractable task... Currently, we are seeing only attempts to introduce means of guaranteed influence on the individual and mass consciousness." [53] He concludes that although the development of a theory of conscientious war—a war of worldview—will turn out to be so promising that ultimately the use of methods of manipulating individual and public consciousness will become as simple as delivering missile and air strikes against troops, it was too early to talk about the existence of such methods. This echoes the idea of a war of consciousness developed by Potekhin and Gromyko. Orlyansky noted the central role of information in such an approach, used simultaneously as a means of influence on the individual and public consciousness and a means of protection against such influence.[54]

One of the most well-known analyses of the central role of information operations in contemporary conflict was detailed in 2013 by Chekinov and Bogdanov, who described the characteristics of what they labeled "new-generation warfare."[55] As discussed earlier, "new-generation warfare" was deemed to be effected principally by nonmilitary means in order to persuade an adversary that any military response would be too costly. Information operations would be central to this: Chekinov and Bogdanov argued that decisive engagements would take place not on the battlefield, but in the information environment as actors struggled for information superiority through the extensive use of propaganda and electronic warfare to disable the adversary's communication, command, and control capabilities. Russia has sought to develop AI technologies to counter the threat from information warfare: former Deputy Defense Minister Borisov stated in March 2018 that artificial intelligence would be necessary for Russia to effectively contest the information environment and win in cyber wars.[56]

Information superiority is thus a central concept for Russian military thinkers.[57] Orlyansky defined information warfare within the context of information superiority, describing it as "measures to capture and maintain information superiority over the enemy (or reduce their information superiority) during the preparation for and during the course of hostilities."[58] Others believed that this definition was too abstract. Donskov and Fomin argue that the essence of information superiority lies in "the presence of a clear advantage...in the efficiency (speed) and quality of information support."[59] Again, there is a significant focus on US thinking and how information superiority is defined in US doctrine. Information superiority is ensured by the quality and quantity of information and the speed with which it can be transferred, thus electronic warfare and disrupting an adversary's ability to receive, transmit, and process information is considered critical. Bogdanov emphasized the importance of information in a 2004 article on the future of warfare, stating that deception and disinformation would be used to conceal preparations for a military operation, as well as the timeframe and scale of mobilization:

The wars and armed conflicts of the past few decades give reason to
believe that it will be impossible to attain strategic and operational
objectives in future wars without achieving superiority over the
adversary in the information sphere.[60]

Measures would comprise a "complex of interconnected and...coordi-
nated...measures conducted through diplomatic channels," as well as
state-controlled and private media outlets, civilian administration, and
military command and control facilities: "appropriate statements by high
ranking political and military officials can be planned for disinformation
purposes."[61] Deception is considered to be an integral component of both
armed conflict and information warfare.[62] Within this, military deception
(*maskirovka*) is considered to be one of the most effective ways of coun-
tering an adversary's ability to penetrate Russia's information systems,
which are described as the most important component in the information
warfare system. *Maskirovka* is defined as a series of interrelated measures
conducted in order to hide troops and objects from an adversary and
mislead them about the presence, location, composition, state, actions,
and intentions of the Russian military. It is conducted at the tactical,
operational, and strategic level, and is frequently aimed at deceiving
foreign intelligence services, which, although not an independent target
of deception, are considered as a "connecting link," a channel for transmitting the required information to decision makers. Thus, *maskirovka* is
closely related to the concept of reflexive control, which aims to induce an
adversary to perform (or not perform) certain actions that are beneficial
to Russia (and its military).[63] Linked to this is the idea of "cognitive
operations" (*kognitivnyie operatsii*), defined by S. S. Sulakshin as attempts
to undermine a country by manipulating its intellectual sphere, feeding
scientists and other experts false information, concepts, and theories.[64]
Another concept related to military deception is the process of "disorganization" (*dezorganizatsiya*) of an adversary's command and control,
measures that are intended to disrupt the function of a command and
control system with the aim of preventing the effective functioning of

forces. It includes information-technical and information-psychological elements, ranging from the destruction of communications systems and electronic warfare to disinformation and cognitive operations.[65]

Reflexive Control

Reflexive control constitutes a specific approach within Russian information confrontation, encompassing the notion of disinformation being deployed to shape an adversary's responses and provoke a specific reaction. It gives prominence to deception and has been described as the "ultimate weapon," an information weapon that is more important in achieving military objectives than traditional firepower, intended to influence an adversary so that he voluntarily takes decisions that are favorable to the initiators of the action. Reflexive control actions can force an adversary into making decisions favorable for the "controlling" side either through intimidation and the threat of damage or by being "lured with advantage," either real or imagined.[66] Thus, disinformation, deception, and *maskirovka* are key methods in the process. Reflexive control can be used both against decisions made by individuals and systems, and it is thought to have been developed by Soviet military strategists in the 1960s, with Vladimir Lefebvre often cited as the "father" of the concept.[67] It is connected to the desire to achieve information superiority, as well as the focus on the "disorganization" of an adversary's forces and targeting decision-making. Time and the element of surprise are considered to be key factors in such actions. An adversary should have very limited time in which to make decisions, and surprise has an unwelcome psychological impact, disrupting plans and forcing premature decisions. M. D. Ionov has set out a number of ways of exercising reflexive control over an adversary:

- *Exerting power pressure/coercion,* including the use (or demonstration) of superior force, psychological attacks, ultimatums, threats of sanctions and risk, military intelligence, provocative military exercises, weapons tests, denying the adversary access to a certain area or isolating it, putting troops on a higher alert status, support for internal destabilizing forces in the enemy rear;

- *Providing false information about the situation*, including military deception, maintaining the secrecy of new types of weapons, changing the methods of operation, and promoting provocative and subversive activities. Other methods include conflict escalation or de-escalation and measures that force an adversary to take retaliatory actions involving a substantial expenditure of forces, assets, and time.

- *Impact on the algorithm of decision-making by the enemy*, including the systematic conduct of exercises through which typical plans are perceived, the publication of a deliberately distorted doctrine, transmitting false data about the situation, and neutralizing operational thinking on the opposing side;

- *Changing the time of decision-making*, which can be achieved through the unexpected outbreak of hostilities, the transfer of information about the situation of a similar conflict to push the adversary to take imprudent action.

Vasily Mikryukov believes that the theory of reflexive control will remain the most important field of research in the near and distant future for both Russian and foreign analysts, and he has suggested that Russia's intervention in Syria was a clear example of the theory being put into practice.[68]

Psychological Warfare from a Russian Perspective

As discussed earlier, Russian military thought divides information confrontation into psychological and technical confrontation. Information-psychological confrontation seeks to influence public consciousness and decision-making within the target in an attempt to undermine its will to resist, institutions, and belief systems—for example, through the use of social media to influence public opinion, divide, and obfuscate.[69] Korotchenko contends that the principal objective of information warfare is the ability to control the "minds of the Russian Federation's population,

[and] undermine its armed forces' moral and combat potential," paving the way for political, economic, and military incursion.[70]

Information-psychological confrontation was a theme discussed during the Cold War, when the Soviets distinguished between the "battle of ideas," based on "truthful information, objective arguments and respect for interstate relations," and psychological warfare, perceived to be "waged with lies, misinformation, ideological subversion, and...direct incitement...against the existing social order."[71] The West was portrayed as engaging in psychological warfare against the USSR and the socialist bloc, with the aim of preventing the spread of socialist ideology. Speaking in 1981 at the 26th Congress of the CPSU (Communist Party of the Soviet Union), Leonid Brezhnev himself accused the West of employing a "whole system of means designed to subvert or soften up the socialist world."[72] The Western media is a specific target of Soviet analysis and anger over the alleged application of psychological warfare techniques against the USSR, including negative reporting about the socialism and justification of US military build-up. Even films were considered to be part of the West's psychological warfare efforts: *Red Dawn*, *Gorky Park*, and other films were described as "whipping up anti-Soviet hysteria."[73]

The conviction that other states were seeking to undermine Russia by means of information and psychological operations endured during the post–Cold War era, when both adversaries and allies were perceived to be engaged in malicious activities designed to exhaust the country's "spiritual and psychological potential," weakening its will to resist and survive.[74] The West was seen to be seeking access to Russia's natural resources, using information pressure and "active measures." This idea has been developed by a number of Russian experts, including M. O. Marichev, I. G. Lobanov, and E. A. Tarasov, who argue that a key trend in contemporary warfare is the so-called "struggle for public consciousness," the desire to control a population's mindset and outlook. The targeting of the public mindset means that the information domain has become critical:

> Modern wars will be characterized by the large-scale use of
> information and other innovative technologies...In the twenty-
> first century, the information sphere has become the main arena
> for developing advanced technologies that have a total impact on
> the consciousness of a population; improving the techniques and
> methods of influencing the mentality of an enemy's population
> has become one of the most important factors in winning a...war.[75]

In order to destabilize a state from within, the widespread use of information tools alongside other nonmilitary means is viewed as a critical
tool for interfering with a population's mindset and manipulating its
"protest potential," a phrase that echoes Gerasimov. M. O. Marichev, I.
G. Lobanov, and E. A. Tarasov characterize this as an "implicit enemy"
and accuse Russia's enemies of using "fifth columnists" and NGOs while
implementing military measures indirectly (in particular through the
use of proxy forces): "in the struggle for mindset," nonmilitary means
are becoming more effective than military ones. According to this view,
information operations erode a population's sense of national and cultural
identity, targeting its traditional values in the name of progress.

There is a widespread belief among Russian military theorists that
strategic competitors, particularly Western states, and domestic opposition groups use information attacks to impose ideas and values "alien
to the Russian mentality" with the aim of dividing Russian society and
undermining national stability, reflecting a "besieged fortress" mindset.
Yury N. Arzamaskin and Oleg V. Kepel accuse Russia's enemies—both
internal and external—of using ideological attacks against the Russian
value system, pointing to the Maidan events in Ukraine in 2013 and 2014
and Belarus in 2020 as examples of this, prompting the need for robust
patriotism as a key line of defense.[76] Zarudnitsky has called for further
work by military theorists within the General Staff on the theory of
information-psychological warfare, arguing that in the foreseeable future
states will continue to face psychological aggression, which he describes
as the "weapon of tomorrow" and a central characteristic of conflict in the

twenty-first century. Modern information and communication technologies have accelerated a change in the forms and methods of psychological warfare, which, in his view, is intended to manipulate society, a country's cultural milieu, and its national mindset.[77] Korotchenko argues that the future of the country and of its armed forces hinges on the capacity to "oppose information aggression in time of peace and war."[78] During the 1990s, even the Russian media was seen by some as posing a threat through the "deliberate distortion" of history and destruction of both national and military traditions. A number of "information-psychological" counteractions were proposed, including the development of a national patriotism to unify the country and provide the basis for all "state-run and independent structures."[79]

Patriotic education continues to be seen as an important element of counteracting information-psychological operations: a robust system of patriotic education that incorporates language, religion, specific Russian "values," history, and historical memory, alongside a readiness to take up arms to protect the state, is judged to provide the best defense against such threats.[80] Bartosh recommends "patriotic education of young people" as a way of reducing the impact of information-psychological operations.[81] The focus on patriotic education as a means of counteracting informational-psychological operations represents a continuation of Soviet-era practices, intended to protect the domestic population from "alien" values and weaken any potential for anti-government protests. J. Lassila argues that the military aspects of patriotic education became increasingly emphasized from 2012 on, in the wake of the protests in Russia during the run-up to the December 2011 Duma elections and the Arab Spring.[82] This militarization increased further in the wake of Russia's annexation of Crimea in 2014 and ongoing conflict with the West.

The issue of patriotic education has been well covered in the military theoretical literature. In addition to detailed articles in *Voennaya Mysl*, the *Voennyi academischekii zhurnal* (Military academic journal) regularly covers the topic and devoted an entire issue to patriotism and military-patriotic education in 2021. A. A. Ozerov declared the imperative of military-patriotic education for Russia's youth, arguing that the values of young people were changing, often "within the framework of liberalism," and they were more inclined to protest:

> Under such conditions, military-patriotic education becomes extremely important, as it will contribute to the development of love for the Motherland, citizenship and the desire to defend the country if necessary.[83]

This echoes a number of articles written by Lutovinov. In 2009 he identified military-patriotic education and the prevention of the spread of "alien" culture to be key nonmilitary measures that could be implemented to strengthen Russia's military security.[84]

Patriotic education is thus considered to be a critical defense against psychological warfare, reflecting a belief that conflict in the twenty-first century is as much about the cognitive domain as the physical. It serves two interlinked purposes: to stimulate internal cohesion around the idea of external threats to Russia (the besieged fortress narrative) and to foster a desire to serve in the armed forces in order to defend the country. In 2018, it was announced that the Armed Forces' Main Military-Political Directorate was being re-created, headed by the former commander of the Western Military District (and frequent contributor to military theoretical debates) Colonel-General Andrei V. Kartapolov.[85] The establishment of a political apparatus within the military echoes Soviet-era structures: it is considered to be the "heir" of the Main Political Directorate of the Soviet Army, itself the successor of the Political Directorate of the Red Army, established in 1918.[86] The new directorate is responsible for all patriotic education in Russia, as well as the military-patriotic youth movement "Yunarmiya," created by Shoigu in 2015.[87] This accentuates the militarization of patriotic education and the importance assigned to it regarding the defense and security of Russia. Kartapolov himself has stated that the directorate was set up to defend the country against the "information war" being waged against it and protect society from "change in the political consciousness," emphasizing that the primary reason for its creation was the "need for information protection of the armed forces...and the development of their belief in the need and importance of serving."[88] Thus, the defense of Russian values and morals is perceived to be as important as military security.

The threat posed to Russia by hostile information operations has been a consistent feature of the country's Military Doctrines (which, as discussed in chapter 1, outline the official view of the character of conflict) since 2000. The Military Doctrine of 2000 categorized the "exacerbation of information confrontation" as a significant determinant of the military-political situation and identified hostile information operations (both technical and psychological) that harmed the country's military security as a principal external threat to national security. The 2010 Military Doctrine marked a significant change to the previous iteration, identifying information warfare as a key characteristic of contemporary conflict. The document noted the growing application of information warfare measures by actors who pursue their strategic objectives without resorting to the use of military force, while simultaneously seeking to shape international support for any future military operation. The 2010 Military Doctrine called for improved information support for the Russian armed forces, along with the development of "forces and resources" for information warfare. The 2015 iteration of the Military Doctrine noted that, although there had been a reduction in the risk of a large-scale war being initiated against Russia, military risks and threats had shifted to the information space and threatened to undermine domestic stability. One of the principal risks to the country's military security is identified as:

> The use of information and communication technologies for military-political purposes to take actions which run counter to international law, being aimed against sovereignty, political independence, territorial integrity of states and posing a threat to international peace, security, global and regional stability.[89]

Subversive information activities directed against the Russian population, particularly the young, with the intention of undermining national values, including historical, spiritual, and patriotic traditions, are also categorized as a key risk. Defining the characteristic features of contemporary military conflicts, the 2015 Military Doctrine outlines the integrated use of both military and nonmilitary means, including political, economic, and information means, together with the widespread use of the "protest potential of the population and special forces," "massive" use of electronic warfare, information and control systems, and simultaneous pressure against an adversary in the global information environment, as well as the land, maritime, air, and space domains. In response, the Russian leadership included information in the list of instruments for protecting the state, placing it on the same level as military, political, and economic instruments.

Russia feels at risk from nonmilitary means of warfare, particularly information confrontation, and a distinctive focus of Russian military thought in recent decades has been on nonmilitary means of achieving strategic objectives, manipulating a state's population and undermining it from within, weakening its will to resist. The information domain and information warfare (which includes cyber operations) have been a central focus, constituting a critical enabler and important means. These means—tools of indirect action—seek to target society, along with the decision-making of a state, thereby undermining its ability to resist an opponent. New technologies and information are perceived to be both critical enablers and key instruments of these nonmilitary approaches, contributing to a significant shift in the character of conflict. There is a deep-rooted concern that indirect actions, including internal destabilization and information warfare, are part of the character of conflict in the twenty-first century, a new means for an adversary to achieve its strategic objectives without the need for a large-scale military invasion. As discussed previously, they are deemed to be part of a new US-led approach to warfare, involving the internal destabilization of adversaries mostly by nonmilitary means. This allows the Kremlin to point to the ostensible threat from the US and the West, which it believes have sought to promote democracy around the world in the controversial pursuit of international security and stability at the expense of state sovereignty and the principle of noninterference in the internal affairs of other states. Although this seems counterintuitive, military capabilities remain critical to hostile nonmilitary measures, underpinning their efficacy. Furthermore, the framing of internal destabilization and so-called "controlled chaos" as an external military threat to Russian national security has facilitated the expansion of defense and security responses to an essentially political issue, allowing the military to justify its involvement and interest in a range of nonmilitary activities, again in the name of "protecting" state sovereignty and regime stability.

NOTES

1. Gorbunov and Bogdanov, "O kharaktere vooruzhennoi bor'byi v XXI veke."
2. Andrei V. Kartapolov, "Uroki voennikh konfliktov, perspecktivyi razvitiya sredstv i sposob ikh vedeniya. Pryamiye i nepryamiye destviya v sovremennikh mezhdunarodnikh konfliktakh," *Vestnik akademii voennikh nauk* 2, no. 51 (2015): 29.
3. Kurochko, "Neklassicheskiye voinyi sovremennoi epokhi," 1–2.
4. Yulia Kozak, "Tuman 'gibridnykh' voin cgushchaetsya nad mirom," *Krasnaya Zvezda* no. 46, April 26, 2019, 8.
5. For more on this see Ofer Fridman, *Russian "Hybrid Warfare": Resurgence and Politicisation* (London: Hurst & Co., 2018).
6. The Brezhnev Doctrine, enunciated in the autumn of 1968, represented the formal manifestation of this view, justifying Soviet intervention in the internal affairs of any country perceived to be threatening the integrity of the socialist bloc.
7. Aleksandr Bartosh, "Kakoi budet strategiya protivostoyaniya v XXI veke," *Nezavisimoe Voennoe Obozrenie* no. 1 (January 11, 2019), https://nvo.ng.ru/nvo/2019-01-11/1_1029_strategy.html.
8. Aleksandr Bartosh, "Vozvrashcheniye bumeranga tsvetnyikh revolyutsii," *Nezavisimoe Voennoe Obozrenie* (April 3, 2021), https://nvo.ng.ru/gpolit/2021-03-04/1_1131_revolutions.html.
9. Basil Liddell Hart, *Strategy: The Indirect Approach*, rev. ed. (London: Faber and Faber Ltd, 1967), 24–25.
10. Sergei G. Chekinov and Sergei A. Bogdanov, "Vliyanie nepriyamykh deistvii na kharakter sovremennoi voini," *Voennaya mysl'* 4, no. 6.
11. Kartapolov, "Uroki voennikh konfliktov, perspecktivyi razvitiya sredstv i sposob ikh vedeniya. Pryamiye i nepryamiye destviya v sovremennikh mezhdunarodnikh konfliktakh," 29.
12. Kartopolov, "Uroki voennikh konfliktov, perspecktivyi razvitiya sredstv i sposob ikh vedeniya. Pryamiye i nepryamiye destviya v sovremennikh mezhdunarodnikh konfliktakh," 30.
13. For example M. A. Artyukh, "Kulturno-informatsionnoye protivoborstvo: istoriya i sovremennost," *Voennaya mysl'* no. 5 (May 2020): 102–111; Aleksandr Bartosh, "Vozvrashcheniye bumeranga tsvetnyikh revolyutsii,"

Nezavisimoe Voennoe Obozrenie, April, 3, 2021, https://nvo.ng.ru/gpolit/ 2021-03-04/1_1131_revolutions.html.

14. For further information see D. C. Earnest and J. N. Rosenau, "Signifying nothing? What complex systems theory can and cannot tell us about global politics" in N. E. Harrison, ed., *Complexity in World Politics: Concepts and Methods of a New Paradigm* (New York: State University of New York Press, 2006); Michael J. Mazarr, *Chaos Theory and U.S. Military Strategy: A "Leapfrog" Strategy for U.S. Defense Policy* (1996), http://www. dodccrp.org/html4/bibliography/comch11.html.

15. Bartosh, "Vozvrashcheniye bumeranga tsvetnyikh revolyutsii."

16. Steven R. Mann, "Chaos Theory and Strategic Art," *The US Army War College Quarterly: Parameters* 22, no. 1 (1992): 66, https://press. armywarcollege.edu/parameters/vol22/iss1/19.

17. The White House, *The National Security Strategy of the United States of America,* September 17, 2002, https://nssarchive.us/wp-content/uploads/ 2020/04/2002.pdf.

18. The State Duma of the Russian Federation, Stenogramma zasedaniya, May 12, 2005 [Transcript of session of May 12, 2005], N 97 (811), http:// transcript.duma.gov.ru/node/1089/#sel=.

19. A. N. Belsky and O. V. Klimenko, "Politicheskie tekhnologii 'tsvyet- nykh revolyutsii': puti i sredstva protivodeystviya," *Voennaya mysl'* no. 9 (September 2014): 3–11. See also G. Filimonov, N. Danyuk, and M. Yurakov, *Perevorot* (St. Petersburg: Piter, 2016).

20. Gene Sharp, *From Dictatorship to Democracy* (London: Serpent's Tail, 2012).

21. Makhmut A. Gareev, *Rossiya v voinakh XX veka.* Transcript of public lecture by General Gareev on March 25, 2004, as part of *Polit.ru*'s Public Lecture series. Published April 1, 2004, http://www.polit.ru/article/20 04/04/01/gareev/.

22. Makhmut A. Gareev, "Obespechenie bezopasnosti stranyi – rabota mno- goplanovaya," *Voenno-promyishlenniyi kur'er* 3, no. 420 (January 4–5, 2012): 25–31.

23. Vladimir Putin, "Byit' cil'nyim: garantii natsional'noi bezopasnosti dlya Rossii," *Rossiiskaya Gazeta* 5708 no. 35 (February 20, 2012): https://rg. ru/2012/02/20/putin-armiya.html.

24. Gerasimov, "Tsennost' nauki v predvidenii."

25. Valery V. Gerasimov, *O roli voennoi cilyi v sovremennyikh konfliktakh.* Speech delivered at III Moscow Conference on International Security

MCIS-2014, May 23, 2014, https://function.mil.ru/news_page/country/
more.htm?id=11929743%40egNews&m=.

26. Belsky and Klimenko, "Politicheskie tekhnologii 'tsvyetnykh revolyutsii': puti i sredstva protivodeystviya," 3–11.

27. Gerasimov, "Razvitie voennoi strategii v sovremmenykh usloviyakh.
Zadachi voennoi nauki," 6–11.

28. See for example Kozak, "Tuman 'gibridnykh' voin cgushchaetsya nad
mirom"; Aleksandr Bartosh, "Agressiya novogo tipa," *Nezavisimoe
Voennoe Obozrenie* (May 18, 2018): https://nvo.ng.ru/vision/2018-0
5-18/1_996_agression.html?; Artyukh, "Kulturno-informatsionnoye
protivoborstvo: istoriya i sovremennost," 102–111.

29. Quoted in Bartosh, "Agressiya novogo tipa."

30. V. Potekhin, "Konstsientral'naya voina," Conscient War, n.p., http://
chomsky.narod.ru/kom56/prop/pkon.htm; Yu Gromyko, "Oruzhie
porazhayushchie soznanie – chto eto takoe?" Moscow School of
Conflictology, https://conflictmanagement.ru/oruzhie-porazhayushhie-
soznanie-chto-eto-takoe.

31. He went on to state that Russia was developing AI technologies to counter
this threat. "Minoboronyi RF: razvitiye iskusstvennogo intellekta nuzhno
dlya uspeshnogo vedeniya kibervoin," *Tass news agency,* March 14, 2018,
https://tass.ru/armiya-i-opk/5028817

32. Gerasimov, "Tsennost' nauki v predvidenii."

33. Zarudnitsky, "Kharakter i soderzhaniye voennyikh konfliktov v sovremennykh usloviyakh i obozrimoi perspective."

34. Aleksandr A. Migunov, "Tendentsii kitaiskoi strategii vedeniya informatsionnoi voinyi," *Voennaya mysl'* no. 11 (2008): 64.

35. Anatoliy A. Novogitsin, "V tsentre vnimaniya – informatsionnaya
bezopasnost" *Krasnaya Zvezda*, February 27, 2009, http://old.redstar.ru/
2009/02/27_02/1_06.html

36. Keir Giles, *Handbook of Russian Information Warfare*, Fellowship Monograph 9, NATO Defense College, NDC Fellowship Monograph Series
(Rome, NATO Defense College, 2016).

37. I. V. Puzenkin and V. V. Mikhailov, "Rol' informatsionno-psikhologich-
eskikh sredstv v obespechenii oboronosposobnosti gosudarstva," *Voennaya mysl'* no. 7 (July 2015): 12.

38. A. Krikunov and A. Korolyov, "Features of Information Warfare of the
Future," *Voennyi Diplomat* no. 1 (2009): 94–103.

39. Aleksandr Bartosh, "Gibridnaya voina stanovitsya novoi formoi mezhgosudarstvennogo protivoborstva," *Nezavisimoe Voennoe Obozrenie,* April 7, 2017, https://nvo.ng.ru/concepts/2017-04-07/1_943_gibryd.html/.

40. Aleksandr N. Limno and M. F. Krysanov, "Informatsionnoe protivoborstvo i maskirovka voisk," *Voennaya mysl'* no. 5 (May 2003): 70.

41. Vyacheslav V. Kruglov, "Novyi podkhod k analizu sovremmennogo protivoborstva," *Voennaya mysl'* no. 12 (2006): 50–61. V. D. Ryabchuk amd V. I. Nichipor, "Filisofiya voinyi i teoriya upravleniya sovremennyim protivoborstvom," *Voennaya mysl'* no. 8 (2007): 65–73.

42. E. N. Dezhin, "Informatsionnaya voina po vzglyadam kitaiskikh voennykh analitikov," *Voennaya mysl'* no. 6 (November 1999).

43. Dezhin.

44. Migunov, "Tendentsii kitaiskoi strategii vedeniya informatsionnoi voinyi," 62–67.

45. A 2020 article detailed Chinese approaches to information warfare, the Chinese perspective on the role and place of information warfare in terms of national security, and the specific measures taken by the country's leadership to strengthen its cyber security. R. A. Polonchuk and T. A. Ganiyev, "View of Chinese military experts on the essence and content of information warfare in the present-day context," *Military Thought* no. 2 (2020): 174–181.

46. Nevertheless, Krikunov and Korolyov assert that the USSR's lack of attention to information warfare contributed to its "defeat" in the Cold War. A. Krikunov and A. Korolyov, "Features of Information Warfare of the Future," *Voennyi Diplomat* no. 1 (2009): 100.

47. Kalinovsky, "Informatsionnaya voina – eto voina?," 59.

48. Sergei A. Komov, "Informatsionnaia bor'ba v sovremennoi voine: voprosy teorii," *Voennaya mysl'* no. 3 (1996): 76–80.

49. M. A. Rodionov, "Forms of Information Warfare," *Military Thought* no. 2 (1998): 84–88.

50. M. A. Rodionov, 84–88.

51. Orlyansky, "Vooruzhennaya i informatsionnaya bor'ba: sushchnost i vzaimosvyaz ponyatii i yavelenii," 46.

52. Vladimir I. Orlyansky, "Informatsionnoye oruzhye i informatsionnaya bor'ba: realnost i domyslyi," *Voennaya mysl'* no. 1 (January 2008): 63–64.

53. Orlyansky, 65.

54. Orlyansky, 68.

55. Sergei G. Chekinov and Sergei A. Bogdanov, "The Nature and Content of New-Generation War," *Voennaya mysl'* no. 4 (2013). This article built

of previous work by the authors. In 2004, Bogdanov published an article on future warfare, asserting that several months before an international conflict breaks out, plans are made for large-scale activities in all forms of confrontation (warfare)—informational, psychological, ideological, diplomatic, and economic—in order to shape favorable military-political and economic conditions for the deployment of military force. Sergei A. Bogdanov, "Warfare of the Future," *Military Thought* no. 1 (2004): 33–38.

56. "Minoboronyi RF: razvitiye iskusstvennogo intellekta nuzhno dlya uspeshnogo vedeniya kibervoin," March 14, 2018, *Tass news agency*, https://tass.ru/armiya-i-opk/5028817.

57. See for example, Yury E. Donskov and V. V. Fomin, "Informatsionnoye prevoskhodstvo: puti realizatsii v operatsiyakh," *Voennaya mysl'* no. 11 (2003): 57–61.

58. Orlyansky, "Vooruzhennaya i informatsionnaya bor'ba: sushchnost i vzaimosvyaz ponyatii i yavelenii," 43.

59. Donskov and Fomin, "Informatsionnoye prevoskhodstvo: puti realizatsii v operatsiyakh," 57.

60. Bogdanov, "Warfare of the Future," 35.

61. Bogdanov, 35.

62. Vladimir I. Orlyansky, "Vooruzhennaya i informatsionnaya bor'ba: sushchnost i vzaimosvyaz ponyatii i yavelenii," *Voennaya mysl'* no. 6 (2002): 42–47.

63. Limno and Krysanov, "Informatsionnoe protivoborstvo i maskirovka voisk," 71.

64. S. S. Sulakshin, "Kognitivnoye oruzhie – novoe pokolenie informatsionnogo oruzhiya" *Vestnik Akademii voennykh nauk* no. 1, 46 (2014): 57–65.

65. For more on disorganisation see Yury E. Donskov, A. L. Morarescu, and V. V. Panasyuk, "K voprosu o dezorganizatsii upravlenii voiskami (silami) i oruzhem," *Voennaya Mysl* no. 8 (August 2017): 19–25; N. A. Kostin, *Teoriya informatsionnoi borby: monografiya* (Moscow: VAGSh, 1996).

66. M. D. Ionov, "On Reflexive Enemy Control in a Military Conflict," *Military Thought* no. 1 (1995).

67. For a detailed examination of the roots of reflexive control and its evolution, see Antii Vasara, "Theory of Reflexive Control. Origins, Evolution and Application in the Framework of Contemporary Russian Military Strategy," *Finnish Defence Studies* 22 Finnish National Defence University, 2020.

68. Vasily Mikryukov, "Nauka pobuzhdat," *Voenno-promyshlennyi kur'er* 629, no. 14, April 13, 2016, https://vpk-news.ru/articles/30204.
69. Migunov, 64.
70. E. G. Korotchenko, "Informatsionno-psikhologicheskoe protivoborstvo v sovremennykh usloviiakh," *Voennaya mysl'* no 1. (1996): 22–28.
71. Y. Solodukhin, "The battle of ideas, or psychological warfare," *International Affairs* 27, no. 10 (1981): 47.
72. Solodukhin. See also A. Leonidov, "The strategy of 'psychological warfare,'" *International Affairs* 5 no. 4 (1959): 30–37.
73. Grigory Oganov, "Behind the scenes of 'psychological warfare,'" *International Affairs* 33, no. 8 (1987): 91.
74. Korotchenko, "Informatsionno-psikhologicheskoe protivoborstvo v sovremennykh usloviiakh."
75. M. O. Marichev, I. G. Lobanov, and E. A. Tarasov, "Bor'ba za mental'nost – trend sovremennoi voiny," *Voennaya Mysl* no. 8 (August 2021): 48–55.
76. Yury N. Arzamaskin and Oleg V. Kepel, "Gosudarstvenno-patriotich-eskaya ideya kak ideologicheskaya osnova voenno-politicheskoi rabo-tyi," *Voennaya Mysl* no. 11 (November 2021): 128–138.
77. Zarudnitsky, "Faktoryi dostizheniya pobedyi v voennykh konfliktakh budushchego," 38.
78. Korotchenko, "Informatsionno-psikhologicheskoe protivoborstvo v sovremennykh usloviiakh."
79. Korotchenko, "Informatsionno-psikhologicheskoe protivoborstvo v sovremennykh usloviiakh."
80. Arzamaskin and Kepel, "Gosudarstvenno-patrioticheskaya ideya kak ideologicheskaya osnova voenno-politicheskoi rabotyi," 128–138.
81. Aleksandr Bartosh, "O gibridnoi agressii i neobkhodimoi oborone," *Nezavisimoe Voennoe Obozrenie*, January 14, 2021, https://nvo.ng.ru/gpolit/2021-01-14/8_1124_challenge.html.
82. J. Lassila, "An Unattainable Ideal: Youth and Patriotism in Russia," in *Nexus of Patriotism and Militarism in Russia: A Quest for Internal Cohesion*, ed. K. Pynnöniemi (Helsinki: Helsinki University Press, 2021), 122.
83. A. A. Ozerov, "Smylsozhizhnennyie osnovy voenno-patrioticheskogo vospitaniya molodezhi," *Voennyi academischekii zhurnal* 1, no. 29 (2021): 50–54.
84. Vladimir I. Lutovinov, "Razvitie i ispolzovanie nevoennykh mer dlya ukrepleniya voennoi bezopasnosti Rossiiskoi Federatsii."
85. "Glavnoye voenno-politicheskoye upravlenie sozdano v Minoborony," July 30, 2018, *Tass news agency*, https://tass.ru/armiya-i-opk/5413977.

86. Aleksei Nikolskii, "V Minoborony sozdano Glavnoye voenno-politicheskoye upravlenie," July 30, 2018, *Vedomosti*, https://www.vedomosti.ru/politics/articles/2018/07/30/776924-glavnoe-upravlenie.
87. "Patrioticheskii zam dlya Shoigu: kto stal novym kuratorom 'Yunarmii,'" *RBC*, July 30, 2018, https://www.rbc.ru/politics/30/07/2018/5b5e13479a79 47a33e086e74. For more on Yunarmiya, see Jonna Alava, "Russia's Young Army: Raising New Generations of Militarized Patriots" in *Nexus of patriotism and militarism in Russia: a quest for internal cohesion*, ed., K. Pynnöniemi (Helsinki: Helsinki University Press, 2021).
88. 'The Military Doctrine of the Russian Federation', approved by the President of the Russian Federation on December 25, 2014, no. Pr-2976.
89. 'The Military Doctrine of the Russian Federation', approved by the President of the Russian Federation on December 25, 2014, no. Pr-2976

CHAPTER 6

ALL AVAILABLE MEANS?

The character of conflict and future war is transformed not just by the weapons being used and the impact of technological change but also by who is doing the fighting. The end of the Cold War witnessed a shift from interstate conflict between national armed forces to a growing role for non-state actors: Mary Kaldor distinguished between "old wars" fought between states by national militaries and "new" wars that comprised an array of combatants, including national armed forces, insurgents, militias, and private military companies.[1] Russian analysis of the trends in contemporary conflict and warfare has focused on how war is being fought, the means (i.e., weapons and systems) being used, and who is perceived to be fighting. A widespread conclusion among Russia's military scientists is that major powers have been "contracting out" the use of force to private companies, prompting a rise in the number of private military forces participating in operations on behalf of a state. This has led to questions about whether states still hold a monopoly on the use of force and whether this been eroded by the participation of violent non-state actors in conflict.

This chapter considers one of the most prominent features of Russia's approach to the use of force in the twenty-first century: the use of proxy, irregular, and private military forces. The use of irregular forces is not a new development for Russia: forces such as the Cossacks have been used for centuries. Furthermore, during the Cold War, the USSR supported proxy forces around the world, and since 1991, Russia has used proxies in pursuit of its objectives across the post-Soviet space. Nevertheless, there has been a gradual evolution in its approach with the emergence of private military and security companies linked to Russia, such as the Wagner Group, which has been increasingly prominent in Ukraine, Syria, and a number of African countries. This chapter examines the roots of this evolution and explores the influence of Western activities, linking it to the evolution in military thought that has occurred in recent decades and subsequent emulation of Western militaries.

NON-STATE ACTORS AND RUSSIAN VIEWS OF THE CHARACTER OF CONFLICT

Russian understanding of the use of proxy forces, including the increasing deployment of private military and security companies[2] and other non-state actors, is connected to their understanding of the character of contemporary conflict, which considers the use of proxy forces, particularly PMSCs, to be a key component of an indirect approach and the West's supposed application of hybrid warfare. In a 2013 article, Gerasimov referred to new types of warfare emerging that were not exclusively military, citing the example of operations in Libya, where private military companies worked closely with opposition forces.[3] Vasily Mikryukov, a member of the Russian Academy of Military Science, has pointed to what he sees as the prevalence of proxy war in contemporary conflict and emphasized the need for Russia to develop "rational behavioral strategies...in preparation for war," stating that following bad strategy is better than no strategy at all:

> The task of Russian military science at present is to study the features of modern proxy wars and to develop theoretical founda-

tions…Comprehending the strategic wisdom of a new war ensures the ultimate superiority of the national strategy.[4]

A number of Russian military theorists have emphasized the increasing importance of non-state actors in contemporary conflicts, including the deployment of PMSCs.[5] Writing in 2016, Aleksandr Kudryavtsev concluded that the world was entering a "difficult era of proxy wars" linked to the emerging concept of hybrid warfare, which in his opinion is being driven partly by the multiplicity of participants in contemporary conflict, from regular troops to paramilitaries and PMSCs.[6] He discerns a clear link between proxy warfare, the role of PMSCs, and contemporary concepts of conflict. Aleksandr Kalistratov has argued that, although contemporary "hybrid wars" include the active involvement of "irregular military formations" and PMSCs, national armed forces continue to play the principal role in security operations, largely because, in his opinion, the significance of military forces, such as militias and paramilitaries, generally turns out to be less than envisaged at the beginning of an operation.[7] In their book on future war, Popov and Khamzatov point out that there have always been "irregular forces" (*irregulyarnyie sili*) on the battlefield; mercenaries and private forces are thus nothing new. However, they divide irregular forces into two distinct groups: active forces, those taking direct part in military operations, and passive ones, including intelligence, logistics, medical units, and guard duties.[8] In their view, private military companies are part of "new-era" conflict, a demonstration of the evolving character of conflict. Others agree with this assessment and urge caution, pointing to the apparent weakening of the state's monopoly on the use of force with a growing number of private paramilitary groups competing with national armed forces.[9] In their analysis of the phenomenon of the privatization of military force, K. Kurilev, E. Martynenko, N. Parkhitko, and D. Stanis contend that one of the principal challenges to international security is the "ever-increasing loss of the state's monopoly on the use of military force," noting that non-state forces are playing a growing role in contemporary conflict.[10]

In his analysis of future war, Kiselyev has pointed to the key role of PMSCs in regime change and hybrid-type wars, which allow states to avoid deploying their armed forces.[11] In his 2021 analysis of the character of modern armed conflict, Zarudnitsky notes the growing role of non-state actors such as PMSCs, arguing that modern technologies have facilitated the emergence of these "new players," putting them on a par with traditional actors on the battlefield—meaning that in the twenty-first century the conduct of hostilities is no longer the exclusive prerogative of states and their armies.[12] Concern about the apparent privatization of conflict was expressed by Tatyana Gracheva, who argued that when proxy forces such as PMSCs are used it eliminates the Clausewitzian trinity of government (reason), the military (violence and enmity), and the people (passion), alienating the population and removing the prospect of victory, making the conflict permanent.[13]

There is little Russian work specifically on Russian PMSCs, either by military theorists or security experts. Work on PMSCs tends to frame them as a Western phenomenon and focuses predominantly on the increasing utilization of PMSCs by Western actors, together with extensive analysis of the implications of this for Russia and appeals for it to develop a similar approach. An exception are two books on the evolution of Russian PMSCs by Vladimir Neelov[14] and another by Ivan P. Konovalov and Oleg V. Valetskii,[15] both published in 2013. Russia's military doctrine, which constitutes an official statement of the Russian view of the character of conflict, referenced PMSCs for the first time in 2014, identifying the participation of "irregular military formations and private military companies" as well as the use of "indirect and asymmetric" methods to be characteristic features of contemporary military conflicts.[16] Neither the 2010 nor the 2000 iterations make any specific mention of PMSCs, but the 2000 Military Doctrine refers to the use of indirect means and the participation of "irregular armed formations" in conflict alongside regular troops, while the 2010 document identifies the "integrated utilisation of military force and forces and resources of a nonmilitary character."[17] The 2000 Military Doctrine also notes that contemporary conflict is characterized by the use of irregular armed formations. The principal focus of the Military Doctrines until 2015 is on state-based threats; although there is some mention of the potential threat from "illegal armed formations," an oblique reference to military groups such as those that had been seen in the North Caucasus, the 2000 and 2010 Military Doctrines concentrate on conventional armed forces and the threat posed by state actors.

PROXY FORCES: AN EVOLVING STRATEGY OF INDIRECT ACTION?

Russia's use of proxy forces could be considered to be part of Vorobyev and Kiselyev's "strategy of indirect actions," which was discussed earlier. Chekinov and Bogdanov returned to the concept of indirect action in 2011, arguing that the increasing use of nonmilitary means to achieve strategic objectives in the contemporary security environment had been driven by the "catastrophic consequences" of conflict involving modern weaponry. According to them, nonmilitary means are related to the concept of indirect actions: both are focused on the defeat of an adversary by indirect, often ambiguous, methods.[18] Both articles refer to the importance of striving to be strong in those areas where an adversary is weak and position strength against weakness, along with the use of surprise. Indirect actions are used to obtain opportunities to ensure the development of the state and contribute to the "weakening and elimination of military dangers and threats." Surprising an adversary deprives him of freedom of action and gives the aggressor the strategic initiative, ideas that play a key role in contemporary Russian military thought.[19] Although information-related means, particularly information-psychological means, are critical to an indirect strategic approach, using proxies is another way of achieving objectives without the use of direct military means, providing a degree of surprise, ambiguity, and deniability that places an actor in a position of strength. Thus, Russia's use of proxies, including PMSCs, should be considered to constitute part of an evolving strategy of indirect action, enabling Moscow to achieve its objectives without necessarily having to deploy military force.

The use of proxy forces by Russia in pursuit of its strategic objectives is nothing new. Writing in the 1920s, Mikhail Tukhachevsky suggested that the Soviet leadership should use partisans as proxy forces to facilitate uprisings and instability elsewhere in the world to undermine adversaries,[20] and the USSR was well-known for using proxies to avoid "hot war" with the US and its allies during the Cold War.[21] Igor Lukes maintains that the Soviets understood the use of force as the most direct instrument of foreign policy and "loaded with risks." Consequently, they sought to avoid provoking the West (particularly the US) into a decisive response, while also camouflaging their use for force through the use of proxies, resulting in a "confused and disunited opponent" and a reduced risk of major conflict with the West.[22] A confused adversary is likely to lose cohesion and be incapable of offering organized resistance. There has been a continuity in Russia's approach in the post–Cold War era, and this description of Soviet logic could well be applied to their contemporary behavior. Following the collapse of the USSR, Russia relied on proxy forces as a cost-effective means of maintaining its interests and influence in the post-Soviet space, predominantly by supporting separatist groups through financial, military, or political means. Covert support was funneled through Russia's 14th Army to separatist groups in Moldova's separatist region of Transnistria during the brief war of 1992, despite it being officially neutral in the conflict. Proxy forces, including armed groups from the North Caucasus, were instrumental in Abkhazia's secessionist conflict with Georgia in 1993. Shamil Basayev, who went on to lead Chechen militants, commanded forces of the Confederation of Mountain Peoples of the Caucasus in Abkhazia from 1992 to 1993, supporting local separatist militias ostensibly with assistance and training from Russian military intelligence.[23]

Moscow has also used proxy forces to deal with internal conflicts, most notably the challenge posed by Chechnya in the 1990s. As discussed in chapter 3, Russia provided covert assistance, both economic and military, to opposition groups and competing factions (including the Interim Council and its forces, led by Beslan Gantemirov) who were opposed to Dzhokhar Dudayev's rule, in an attempt to critically destabilize the regime without officially deploying Russian troops. This culminated in a failed intervention in November 1994, when proxy forces launched an abortive attack against the Chechen capital Grozny, supported by Russian heavy equipment, including tanks, armored vehicles and aircraft, and troops. Moscow continued to deny any involvement until Russian media outlets published evidence from Russian officers, who claimed to have signed contracts with the FSK "to take part in a secret military operation" in Chechnya.[24] This strategy of seeking to exploit divisions between different factions to destabilize was revived during the Second Chechen War and played a key role in Russia's approach to counterinsurgency, most notably the policy of "Chechenization," implemented in 2000. Putin imposed direct rule over the republic and installed a pro-Moscow administration led by Akhmad Kadyrov, a mufti and former insurgent, who had become increasingly concerned about the growing radicalization of many Chechen militant groups and had consequently changed sides. The policy of Chechenization gradually brought an end to major violence, with large-scale conflict in Chechnya gradually ending from 2003 to 2004. Still, the policy was successful partly because it devolved responsibility for brute force down to the Chechen level: the Kadyrov regime employed brutal methods to subdue the population, and the policy of Chechenization has, to some extent, led to increased conflict between Chechens.[25]

As relations with the West deteriorated during the 2000s, the use of proxies again offered reduced risk in terms of intervention, reflecting Soviet-era practices. The International Commission that investigated the August 2008 conflict between Russia and Georgia described it as an amalgamation of an interstate and intrastate conflict, involving military action between regular armed forces, alongside action by militias and irregular armed groups.[26] South Ossetian and Abkhazian militias (*opolchentsy*) supported Russian troops, and there were also reports of

Chechens and Cossacks taking part in the conflict. According to Human Rights Watch, South Ossetian "forces" included servicemen from the South Ossetian Ministry of Defense and Emergencies and the South Ossetian Ministry of Internal Affairs, as well as riot police and volunteers. Distinguishing between the different groups was difficult, with locals referring to most fighters as *opolchentsy*.[27] Close cooperation with militia groups endowed the Russian forces with a significant degree of flexibility, both operationally and diplomatically, as well as ambiguity and therefore deniability. This proved particularly important when allegations of abuse emerged, a situation that has been repeated in Ukraine since February 2022.[28]

The use of proxy forces has been a hallmark of Russia's intervention in Ukraine since 2014, with numerous reports that the Kremlin has been providing significant support to local militias in the Donbas, including groups such as the Vostok Battalion, which was reported to contain both Russian and Ukrainian fighters.[29] Sergey Sukhankin has investigated Russia's reliance on non-ethnic Russian volunteers (including Ukrainians, Belarusians, Georgians, Chechens, and Serbs) in the Donbas conflict, arguing that Moscow is keen to emphasize their role as "defenders of the Russian state" by resuming policies practiced by both the Soviet and Tsarist regimes.[30] The use of local militias has facilitated military operations, enhancing local knowledge and popular support. Moscow's reliance on proxy forces (including PMSCs, discussed later) has also enabled it to maintain a facade of plausible deniability vis-à-vis its involvement in Ukraine, in spite of overwhelming evidence to the contrary. Unsurprisingly, given the Kremlin's apparent desire to retain a level of deniability and ambiguity around its involvement in Ukraine prior to its full-scale invasion in 2022, there is very little military theoretical writing on proxies in Ukraine. What open-source analysis there is focuses on alleged Western involvement. One report by a member of the Academy of Military Science identified proxy war taking place in eastern Ukraine, driven by the support of oligarchs and the US for organizations such as Pravyi Sektor (Right Sector) and Svoboda (Freedom), with the intention of drawing Russia into the conflict.[31] Russia has also found itself on the other side of a proxy war in Ukraine, with Chechens (who had left Russia) fighting against pro-Russian separatist and militia groups in support of Ukraine. There were at least two battalions comprised of Chechens fighting for Ukraine, reportedly at the invitation of Kyiv: the Dzhokhar Dudayev peacekeeping battalion and the Sheikh Mansur peacekeeping battalion, both established by Isa Munaev, a veteran of the post-Soviet Chechen Wars, who is reported as stating that the "fight of the Ukrainian people against imperial Russia is part of our common struggle for the decolonization of the Caucasus."[32]

Russia's intervention in Syria, portrayed by the Kremlin as a Western-style "non-contact" expeditionary operation carried out by the Aerospace Forces, has relied heavily on the use of proxy forces: local and Iranian-backed militias such as Hezbollah were responsible for ground operations while Russia provided close air support. This approach has not been without problems, and Gerasimov has complained about significant challenges in organizing effective cooperation between the Aerospace Forces and proxy forces.[33] In 2016, the Center for Reconciliation of Opposing Sides was set up at the Hmeimim base to manage the ceasefire, the evacuation of civilians, and the provision of humanitarian aid. The commander in charge of the Russian group of forces in Syria is responsible for the Center, which also appears to facilitate coordination between Russian forces and the wide range of other pro-Syrian forces involved in the conflict.[34]

The Syria campaign demonstrates the evolution of military and strategic thought with regard to use of proxy forces and their utility in terms of facilitating Russian strategic objectives without incurring the full costs, both financial and human, of war. Proxy forces tend to have good local knowledge of both the physical and human terrain, bestowing military advantage on the actor on whose behalf they are acting. Russia, like other states, has learned the harsh lessons from its post–Cold War (and Soviet) operational experience, in particular, the need to have public support for military interventions. Russian society, like many others, has become increasingly casualty-averse, reflecting global trends. Russia had also become increasingly aware of the reputational risks associated with overseas military interventions. A 2019 poll by the Levada Center on Russia's intervention in Syria found that although a majority of those questioned approved of Russian involvement (51 percent), 55 percent also believed that Russia must complete its military operation there (up from 49 percent in 2017). Furthermore, an increasing number expressed concern that the situation could escalate into a "new Afghanistan" for Russia.[35] Neelov notes that states have increasingly resorted to the use of PMSCs, both because of the changing character of war and, more importantly, societal reaction to losses. Russia is not immune to this and, in his view, the experiences of Afghanistan and Chechnya mean that the Russian population now reacts badly to the participation of the armed forces in conflict.[36] This was emphasized by the Russian leadership's efforts to hide the scale of the country's losses in Ukraine in the wake of its February 2022 invasion.

The Russian example is consistent with research that has challenged the prevailing view that the use of proxy warfare was a common instrument during the Cold War era but had become less common since 1989. Proxy war has remained an important issue, driven by the reluctance of states to deploy their armed forces, increasingly casualty-averse domestic populations, and the pursuit of more cost-effective means of achieving strategic objectives.[37] In a 2003 article, Philip Bobbitt argues that proxies offer states an "economic alternative" to expensive standing armed forces. The spiraling costs of maintaining and equipping national defense forces during the years of global financial crisis and recession have encouraged states to look for more cost-effective alternatives. According to one scholar, contemporary proxy wars have become "arm's-length 'effects-based operations,'" which facilitate the realization of strategic objectives without the risks of anticipated consequences, such as conflict escalation or heavy financial costs—all without a state having to directly commit military forces of its own.[38]

GROWING RELIANCE ON PRIVATE MILITARY COMPANIES?

A central feature of the contemporary use of proxy forces by states is the growing employment of PMSCs: during the post–Cold War era, Western states became increasingly reliant on contractors to reinforce their militaries. Andrew Mumford has characterized the increasing reliance of states on PMSCs as a defining hallmark of the security policy of Western states in the twenty-first century, citing the example of Iraq where, by 2008, there were nearly two-hundred thousand PMC contractors, exceeding the number of coalition troops.[39] The pursuit of economic efficiencies provide one explanation for this: there are lower costs associated with PMSCs than the deployment of national military forces —for example, states do not have to provide redundancy or pensions for private contractors. The US example of depending on PMSCs to sustain concurrent, long-term operations in Afghanistan and Iraq was watched closely by Russian analysts; there was considerable interest in the case of Blackwater (now known as Academi) and other Western PMSCs.[40] Some estimates suggest that contractors comprised as much as 50 percent of the US military contingent in these two operations.[41] Russia's Defense Minister, Sergei Shoigu, has pointed to the "gigantic" sums of money that the US spends on PMSCs, noting that Russia would never seek to compete with the US military budget.[42] As discussed previously, Russian military thought has been shaped to a significant degree by international operations and interventions, particularly those conducted by Western states. There were a number of articles in the Russian press from the early to mid-2000s examining the American use of PMSCs in Iraq and elsewhere.[43] This has permeated Russian military science, and a report in *Krasnaya Zvezda* on the activities of the Academy of Military Sciences in 2008 notes:

> It is also very important that scientists finally paid attention to the study of such a dangerous and new phenomenon as the role of private security and other military corporations, the number of which begins to exceed the number of regular armed forces in some states. In particular, a working group led by A. I. Nikitin developed

for the UN a draft law 'On Counteraction to Mercenarism' and a draft international convention on the regulation of the activities of private military companies.

Some Russian analysts have pointed to the phenomenon of "double" proxy wars, which they perceive to be becoming more prominent in the contemporary strategic environment. Mikryukov defines a double proxy war as one in which PMSCs or multinational corporations are involved and act as a third party "behind which a state tries to conceal its direct participation and involvement."[44] Major General Aleksandr I. Vladimirov has argued that states are increasingly losing national control over the development of multinational corporations, which he considers to have economic and financial capabilities commensurate with the power of states, enabling them to establish private military and security companies. He expressed his fear that mercenaries could become the principal force of a global multinational alliance, capable of demonstrating "overwhelming real power superiority over state structures...and national armed forces" and inflicting overwhelming defeat against states.[45] This somewhat apocalyptic view emphasizes growing concern about the privatization of force and the implications for national power.

As discussed earlier, Russian understanding of PMSCs and the use of proxies is connected to conceptualization of the character of contemporary conflict. A 2009 report in *Flag Rodiny*, the newspaper of the Russian Black Sea Fleet, included a damning assessment of the growing prevalence of PMSCs and mercenaries. The author refers to the "golden age" of PMSCs associated with the growth in local and global conflicts, an "industry of war," stating that "Iraq and Afghanistan changed everything": American and European companies are deemed to have sent thousands of "alternative armies" to Iraq and Afghanistan, and an in-depth analysis of the Blackwater and Zi PMSCs is provided.[46] Similarly, a 2010 article in *Krasnaya Zvezda* by Aleksandr Mikhailenko[47] examined the growth in PMSCs around the world. Its primary focus was Western ones, but the article also notes the growing presence of Chinese PMSCs, which were seen to be expanding rapidly, particularly in African countries, where they provide security for Chinese oil and gas companies. Mikhailenko set out a variety of reasons for the rapid rise in both the number and employment of PMSCs, including globalization, privatization, outsourcing, and geopolitics. In his view, a number of countries are seeking to bolster their influence around the world through the use of PMSCs; Russia however, "missed the moment" and is almost completely absent from the PMC market, putting it in a weak position vis-à-vis other states.[48] Military expert Aleksandr V. Khomutov has warned that the domains of confrontation in a military conflict are expanding, leading to a growth in the services provided by PMSCs. In his view, this means that there are foreign PMSCs conducting "psychological and cyber operations" against adversaries, including Russia.[49]

There has also been much Russian attention focused on the apparent involvement of Western PMSCs in Georgia prior to the 2008 Russo-Georgian War. Konovalov and Valetskii maintain that the presence of Western PMSCs in the South Caucasus state became a "serious security problem" for Russia and point to allegations that the American firm Military Professional Resources Inc., which had been contracted to prepare Georgian troops for operations in Iraq and Afghanistan, was actually there to help retake Tskhinvali.[50] Olga Sibileva details the supposed involvement of a number of Western PMSCs in Georgia, including the Israeli firm Defensive Shield, and maintains that foreign PMSCs have been active in eastern Ukraine since 2014.[51] As noted above, aside from the work of investigative media such as *Fontanka*, there is very little work by Russian analysts on Russian PMSCs. The work of K. Kurilev, E. Martynenko, N. Parkhitko, and D. Stanis, on PMSCs, which focuses on Ukraine, makes no mention of any Russian PMSCs such as Wagner, focusing their attention instead on the role of Ukrainian PMSCs in the ongoing conflict, as well as the supposed participation of Western PMSCs such as Greystone and Academi, along with groups such as SBS Othago (a Polish firm).[52]

Mikhailenko argues that the activity of private companies in the military sphere has become a reality and needs to be taken into account both in terms of foreign and security policy planning. In his view, they have become an instrument of foreign policy, enabling states to take deniable action and evade legal responsibility, as well as providing "cannon fodder" for Western countries who are reluctant to incur casualties amongst their armed forces deployed on operations in Afghanistan and Iraq. Combat losses amongst the armed forces undermine troop morale and domestic support for overseas interventions; by contrast, contractor casualties tend to go unreported.[53] Sibileva argues that states benefit from the use of PMSCs as a "new form of warfare." According to his analysis, private military forces can be assigned the state's "dirtiest" work, from overthrowing legitimate governments to assassinating political and public figures, in the knowledge that the use of PMSCs provides a flexible, non-attributable force, offering the state deniability and ambiguity.[54] This is certainly reflective of the way that Russian PMSCs have been utilized. Belokon, A. V. Biytyev, and L. A. Smirnova surmise that the principal advantages of PMSCs include efficiency, efficacy and financial advantage: "it is often more profitable [for a state] to sign a contract with a private company...than to send in troops."[55]

PMSCs only really came to prominence in Russia at the beginning of the twenty-first century. This lends weight to the argument that their evolution was influenced by observation of the Western experience and concern that the character of conflict was changing: during the first decade of the twenty-first century, Western PMSCs were very active across the globe, from Iraq to Afghanistan and Libya. Russia's apparent emulation of the Western approach to war and conflict, as discussed in previous chapters, is likely to have been a key driver of the shift toward an increased utilization of PMSCs, along with the economic efficiencies (and profits), flexibility, political deniability, and lack of public accountability that PMSCs offer. Russia's former CGS Nikolai Makarov is reported to have encouraged the use of PMSCs for "delicate missions abroad...to avoid the humiliation of 2004," and other military analysts and experts have suggested that the use of PMSCs would help overcome the challenges related to conscription faced by the Russian military.[56]

Private security companies were legalized in Russia in the 1990s, in order to enable their participation in the protection of energy companies and their facilities. For example, in 1998 the gas giant Gazprom established its own private security company, Gazprom Okhrana, to provide security at its facilities as well as those of its subsidiaries.[57] Russian security companies were contracted to provide protection and guard services to energy facilities in Iraq in the mid-2000s, and Mikhailenko has drawn attention to the presence of Russian contractors there, providing security for the Russian Engineering Company. The contractors were employees of the Antiterror Orel Group, a private security company that was a precursor to Wagner and comprised military and special forces veterans.[58] Groups such as the Moran Security Group provided maritime security and anti-piracy protection in the Straits of Hormuz, as well as training for local forces in the Central African Republic, Nigeria, and Kenya.[59] Other Russian groups to have been involved similar activities include the RSB Group and Tiger Top Rent Security.[60]

There has been some political debate on the issue in Russia. In 2012, during his time as prime minister, Putin expressed his support for the creation of a system of private military companies in Russia that could provide services such as the protection of facilities and infrastructure and the training of foreign military personnel abroad without state participation. Speaking in the Duma in response to a question about the possibility of using Russian PMSCs as an instrument of influence abroad, Putin described PMSCs as a "tool for realising national interests without the direct intervention of the state."[61] In early 2018, the Duma considered draft legislation that would have legalized the creation and activities of private military companies as limited liability companies, assigning licensing of them to the Ministry of Defense. It would also have permitted PMSCs to create subsidiaries on the territory of other states.[62] Nevertheless, despite Putin's apparently positive attitude toward

PMSCs in 2012, the government made its opposition to the 2018 initiative clear, arguing that the provisions of the bill defining the activities of private military companies contradicted Article 13 of the constitution, which prohibits the creation and operation of public associations that are intended to create armed formations.[63] It also pointed to Article 71, which states that issues of defense and security, war and peace, foreign policy, and international relations are the prerogative of the state, and therefore, private companies cannot be involved. A wide range of security agencies and ministries also opposed the draft bill, including the Ministry of Defense, the Ministry of Foreign Affairs, the FSB, the SVR, the Rosgvardiya (the National Guard), the Prosecutor General, and the Ministry of Justice.[64]

The 2018 debate in Duma regarding the legalization of PMSCs reflects the growing role and visibility of such groups in Russia. Although there is little written on PMSCs (or the 2018 draft legislation) in Russian military theoretical journals, there have been a few articles. A 2018 article by Mikhail Gol'dreer in *Arsenal Otechestva* was very critical of the Russian government's denial of PMSCs, describing it as "insulting and naïve." Gol'dreer also condemned the state's rejection of draft legislation on PMSCs, suggesting that, according to the official justification given for the government's opposition, the use of any private companies to supply troops in Syria (such as charter companies) would theoretically be impossible.[65] Certainly, the official position is at odds with a wide range of overwhelming evidence of Russian PMSC activity in a number of territories outside of Russia. Syria has constituted a critical testing ground for the Russian use of PMSCs in a conflict situation. As Russian Defense Minister Sergei Shoigu has stated: "In Syria, we had to learn to fight in a new way. And we have learnt this."[66]

The Wagner Group is the most well-known of the Russian private companies apparently involved in the provision of military contractors. According to the *Fontanka* investigative website, Wagner was formed on the basis of the Slavonic Corps and has been active in Ukraine (in support of pro-Russian separatists), Syria, and Libya, as well as the in Central African Republic, Mozambique, and Mali. Wagner operatives were reportedly among the "little green men" who seized Crimea in 2014 and then participated in the conflict in eastern Ukraine, with groups taking part in the battle for the Donetsk Airport and the assault on Debaltseve, as well as in black operations and assassinations.[67] Although it is not directly linked to the Russian state, personnel from the group are reported to have received military honors and medals, including the Order of Courage, awarded posthumously to a number of men apparently killed while working for Wagner.[68] Its training base is located in Molkino in Krasnodar *krai*, adjacent to a GRU special forces (*spetsnaz*) camp; moreover, its original commander, Dmitry Utkin, is a former GRU officer, suggesting close ties to the Russian security services. Financing of the group is unclear, with one expert suggesting that the only probable source of funding, given the high costs involved, could be "sponsorship" by an oligarch.[69] Yevgeny Prigozhin, who has close ties to the Kremlin and Putin, has long be linked to Wagner.

There were reports in 2019 that as many as four-hundred Russian military contractors linked to the Wagner Group were in Venezuela to protect President Nicolas Maduro in the face of opposition protests backed by the US.[70] In 2021 Russia's Foreign Minister Sergei Lavrov announced during a session of the UN General Assembly that the authorities in Mali had "turned to a private military company from Russia" to help in the fight against terrorism following a decision by France to reduce its involvement in the country. Lavrov was careful to stress that although PMSCs operate in the region legally, they had nothing to do with the Russian government. The involvement of the Russian PMSC, thought to be Wagner, was framed as a response to a request from the African country, which turned to Russia after being let down by the West, in this case France, which "had promised to help eradicate terrorism."[71] Several thousand Wagner mercenaries were deployed to Ukraine in spring 2022 as it became clear that the invasion was not going well for the Russian armed forces.

Putin acknowledged the presence of Russian military contractors in Libya during a press conference with German chancellor Angela Merkel in early 2020, stating that they did not "represent the interests of the Russian state" nor did they receive money from it.[72] Later that year, when asked about the detention of a number of Russians in Belarus supposedly linked to Wagner, the president's spokesman Dmitry Peskov emphasized that Russian legislation does not allow for the existence of PMSCs, only private security companies.[73] Groups such as Wagner provide the Russian government with a deniable security force: the continuing illegality of such groups in Russia enable the government to point to their illegality and claim ignorance. This deniability has been particularly important when Wagner has deployed in the post-Soviet space, for example in Crimea and eastern Ukraine in 2014, and then in Belarus in 2020. PMSCs are cheaper for a state in financial and political terms: they are deniable, provide a state with the cover of ambiguity, and avoid official military casualties and the subsequent public concerns, all while acting as a force multiplier.

Russia's use of proxy forces is not new; proxy forces have traditionally been used to avoid direct confrontation with larger powers, and this continues to be the case. What is new is its increasing use of and reliance on PMSCs in an apparent emulation of the Western experience, with Russia seemingly employing them as a cost-effective force multiplier. The growing prevalence of Russian PMSCs fits with their evolving understanding of the character of conflict in the twenty-first century. As discussed earlier, much of the work by influential military thinkers such as Kiselyev, Bogdanov, Chekinov, Zarudnitsky, and other writers over the past decades has focused on Western operations in Iraq, Afghanistan, and elsewhere and has highlighted the growing use of PMSCs by states, particularly the US. This links to a common Russian view that the role of non-state actors, including business (linked to the PMSC issue) in conflict is growing. This is nothing new nor groundbreaking in terms of conclusions: numerous scholars and analysts have pointed to the growing role for non-state actors in conflict and war in the post–Cold

War era. But it is important to understand the context for the increasing prevalence of Russian PMSCs: the emulation of Western behaviors, in particular the outsourcing of risk and cost to private companies. PMSCs afford the Russian government a very useful degree of ambiguity and deniability, which is amplified by the fact that they officially remain illegal in Russia. This allows Russia to avoid becoming formally involved in a conflict and also avoid official casualties in the state's armed forces (avoiding political, economic, financial, and societal costs). The risk is borne by private companies and the market, rather than the political leadership in the Kremlin. Russia's growing use of PMSCs and proxy forces reflects global trends: it has not just been watching the West but also Turkey, the situation in Yemen and Saudi Arabia's involvement, and China. Russia's use of PMSCs as a flexible, unattributable tool of foreign policy is likely to continue to evolve; it is still at an early stage of development and deployment. Whereas Russian military thought has focused on—and been in response to—Western interventions and practices in recent decades, moving forward Russia may seek to emulate the Chinese example. China legalized PMSCs in 2009, and they are now being used to provide security for the Belt and Road Initiative; use of the People's Liberation Army is not considered to be an option because of the Chinese emphasis on non-interference.[74]

NOTES

1. Mary Kaldor, *New and Old Wars: Organised Violence in a Global Era* (Cambridge: Polity, 2021).
2. There are a range of terms used, including private military companies, private security contractors, and private military and security companies. The term "private military and security company" is used here to reflect the range of activities that such companies conduct, from the provision of military services, logistics and consultancy to security activities such as guarding and risk management. For a detailed examination of the private military industry see Peter W. Singer, *Corporate Warriors: The Rise of the Privatised Military Industry* (London: Cornell University Press, 2004).
3. Gerasimov, "Tsennost' nauki v predvidenii."
4. Vasily Mikryukov, "Povoyuite za menya," *Voenno-promyshlennyi kur'er,* October 5, 2015, https://vpk-news.ru/articles/27400.
5. Vladimir Mokhov and Dmitry Evstafiev, "Myi vedem bitvu za dushi nashykh lyudei," *Armeiskii sbornik* no. 7 (July 2021): 6. Other analysts agree with this assessment; see, for example, Nikolai Poroskov, "Boi netraditsionnoi orientatsii," *Armeiskii sbornik* no. 7 (July 2020): 5–8.
6. Aleksandr Kudryavtsev, "Voennyie konfliktyi tret'yego tyisyacheletiya," *Nezavisimoe Voennoe Obozrenie* no. 13, April 8, 2016, 6.
7. Aleksandr Kalistratov, "Voina i sovremennost," *Armeiskii sbornik* no. 7, (July 2017): 13.
8. Popov and Khamzatov, *Voina budushchego: kontseptualnye osnovy i prakticheskie vyvody,* 251.
9. O. P. Sibilyeva, "Deyatyelnost chastnikh voenniykh kompanii v sovremennikh vooruzhenniykh konfliktakh kak viyzov mezhdunarodnomu gumanitarnomu pravu," *Voennaya mysl'* no. 7 (July 2016): 49.
10. K. Kurilev, E. Martynenko, N. Parkhitko, and D. Stanis, "Fenomen chastnykh voennykh kompanii v voenno-silovoi politike gosurdarstve v XXI v.," *Vestnik mezhdunarodnykh organizatsii* 12, no. 4 (2017): 130–131.
11. Valery A. Kiselyev, "Gibridnaya voina kak novyi tip voiny budushchego," *Armeiskii Sbornik* no. 12 (December 2015): 13; Valery A. Kiselyev and Alexei N. Kostenko, "Kibervoina kak osnova gibridnoi operatsii," *Armeiskii Sbornik* no. 11 (November 2015): 3–6.

12. Zarudnitsky, "Kharakter i soderzhaniye voennyikh konfliktov v sovre-mennykh usloviyakh i obozrimoi perspective," 38.

13. Tatyana Gracheva, "Global'noye chernoye morye," *Voenno-promyshlennyi kur'er* 653, no. 38, October 5, 2016, https://vpk-news.ru/sites/default/files/pdf/VPK_38_653.pdf.

14. A former FSB agent, Neelov was an expert in PMCs and contemporary conflict. His links with a German organization and the "transfer of data abroad" saw him being found guilty of treason by a St. Petersburg court in 2020 and sentenced to seven years in prison. "Byivshii operativnik FSB osuzhden v Peterburge za gosudarstvennuyo izmenu," *Fontanka,* July 2, 2020, https://www.fontanka.ru/2020/07/02/69345721/.

15. Ivan P. Konovalov and Oleg V. Valetskii, *Evoliutsia chastnykh voen-nykh kompanii* (Moscow: Pushkino Tsentr Strategicheskoi Koniunktury, 2013).

16. The Military Doctrine of the Russian Federation. Approved by the President of the Russian Federation on December 25, 2014, No. Pr-2976. For a Russian definition of PMSCs see Nikolai N. Tyutyunnikov, *Voennaya mysl' v terminakh i opredeleniyakh v trekh tomakh. Tom 1: Vooruzhennye sily Rossiiskoi Federatsii* (Moscow: Pero, 2018), 68–70.

17. The Military Doctrine of the Russian Federation. Approved by the President of the Russian Federation on April 21, 2000; The Military Doctrine of the Russian Federation. Approved by Russian Federation presidential edict on February 5, 2010.

18. Chekinov and Bogdanov, "Vliyanie nepriyamykh deistvii na kharakter sovremennoi voini," 4.

19. Liddell Hart, *Strategy: The Indirect Approach.*

20. Raymond L. Garthoff, *How Russia Makes War: Soviet Military Doctrine* (London: G. Allen & Unwin, 1954), 392.

21. See for example Roger E. Kanet, "The Superpower Quest for Empire: The Cold War and Soviet Support for 'Wars of National Liberation,'" *Cold War History* 6, no. 3 (2006): 331–352; Richard E. Bissell, "Soviet use of proxies in the third world: The case of Yemen," *Soviet Studies* 30, no. 1 (1978): 87–106; Igor Lukes, "Great expectations and lost illusions: Soviet use of Eastern European proxies in the Third World," *International Journal of Intelligence and Counter Intelligence* 3, no. 1 (1989): 1–13.

22. Lukes, "Great expectations and lost illusions," 10.

23. "Ex-FSB Head Says Shamyl Basayev Cooperated with Russian Intelligence in Abkhazia," *Civil Georgia,* July 14, 2020, https://civil.ge/archives/359380.

24. German *Russia's Chechen War,* 122.

25. See Tracey German, "Russia-Chechnya: from National Liberation to Deterrence Stability" in *Deterring Terrorism: A Model for Strategy Deterrence*, ed. Elli Lieberman, (New York: Routledge, 2018), 159–180. See also Brian G. Williams, "Fighting with a Double-Edged Sword? Proxy Militias in Iraq, Afghanistan, Bosnia and Chechnya" in *Making Sense of Proxy Wars: States, Surrogates and the Use of Force*, ed. Michael A Innes (Washington DC: Potomac Books, 2012).
26. The Independent International Fact-Finding Mission on the Conflict in Georgia, *Report*, September 2009, https://www.echr.coe.int/Documents/HUDOC_38263_08_Annexes_ENG.pdf, 36.
27. Human Rights Watch, "Up in Flames—Humanitarian Law Violations and Civilian Victims in the Conflict over South Ossetia," *Human Rights Watch*, January 23, 2009, https://www.hrw.org/report/2009/01/23/flames/humanitarian-law-violations-and-civilian-victims-conflict-over-south.
28. The Parliamentary Assembly of the Council of Europe expressed concern about what it described as credible reports of acts of ethnic cleansing committed by militia groups in South Ossetia, which Russian forces failed to stop. Parliamentary Assembly of the Council of Europe, "The Consequences of the War between Georgia and Russia," *Resolution 1633*, adopted on October 2, 2008. Para. 13.
29. Alec Luhn, "Volunteers or Paid Fighters? The Vostok Battalion Looms Large in War withKiev," *The Guardian*, June 6, 2014, https://www.theguardian.com/world/2014/jun/06/the-vostok-battalion-shaping-the-eastern-ukraine-conflict.
30. Sergey Sukhankin, "Foreign Mercenaries, Irregulars and 'Volunteers': Non-Russians in Russia's Wars," in *War by other means: Russia's use of private military contractors at home and abroad*, Special Project, The Jamestown Foundation, October 9, 2019, https://jamestown.org/program/foreign-mercenaries-irregulars-and-volunteers-non-russians-in-russias-wars/.
31. Mikryukov, "Povoyuite za menya," 10.
32. Mairbek Vatchagaev, "Two Battalions of Chechens Now Fighting the Russians in Ukraine," *Eurasia Daily Monitor* 11, no. 199 (November 7, 2014), https://jamestown.org/program/two-battalions-of-chechens-now-fighting-the-russians-in-ukraine-2/#.VNUXLi5RKHg.
33. NewsFront information agency, "Intervyu nachalnika Genshtaba VS RF Gerasimova ob itogakh operatsii VS RF v Siriii i o dalneishikh perspektivakh siriiskoi voinyi."
34. Tikhonov, "Siriiskaya proverka boem."

35. Levada Center, "Sobyitiya v Sirii," Levada Centre, May 6, 2019, https://www.levada.ru/2019/05/06/sobytiya-v-sirii/.
36. Vladimir Neelov, *Chastnyie voennyie kompanii v Rossii: opit i perspektivyi ispol'zovaniya* (St. Petersburg, 2013), 51.
37. Andrew Mumford, "Proxy Warfare and the Future of Conflict," *The RUSI Journal* 158, no. 2 (April/May 2013): 40–46.
38. Mumford, 45
39. Mumford, 42. Private contractors constituted around 57 percent of the total number of personnel deployed.
40. *Fontanka*, a Russian investigative online news site, has conducted a number of analyses on both Blackwater and PMCs in general. For example Denis Korotkov, "Kukhnya chastnoi armii," *Fontanka*, June 9, 2016, https://www.fontanka.ru/2016/06/09/070/. See also Dmitry Novik, "Chernyie Vody," September 15, 2020, https://tjournal.ru/stories/2117 58-chernye-vody.
41. Sean McFate, *The Modern Mercenary* (Oxford: Oxford University Press: 2014), 75.
42. Rostovskii, "Sergei Shoigu rasskazal kak spasal Rossiiskuyu Armiyu," 12.
43. Sibilyeva, "Deyatyelnost chastnikh voenniykh kompanii v sovre-mennikh vooruzhenniykh konfliktakh kak viyzov mezhdunarodnomu gumanitarnomu pravu," 49–61; S. P. Belokon, A. V. Biytyev, and L. A. Smirnova, "Chastniye voennie kompanii: mirovoi istoricheskiy opiyt i sovremenniye perspektiviy dlya Rossii," *Voennaya Mysl* no. 1 (January 2015): 60–66; Mikryukov, "Povoyuite za menya," 10.
44. Mikryukov, , "Povoyuite za menya," 10.
45. Vladimirov, *Osnovyi obshchei teorii voinyi v 3 chastyakh. Chast II: teoriya natsional'noi strategii: osnovyi teorii, praktiki i iskusstva upraveleniya gosudarstvom* (Moscow: Universitet Sinergiya, 2018): 822–823.
46. Vadim Mamlyga, "Zametki voennogo obozrevatelya. Korporatsiya 'voina', ili plokhie parni v chernykh shlyapakh," *Flag Rodiny* no. 113 (October 9, 2010): 5
47. Mikhailenko is Professor of the Department of National Security of the Russian Academy of State Service under the President of the Russian Federation.
48. Aleksandr Mikhailenko, "Voennyie uslugi...v chastnom poryadke," *Krasnaya Zvezda*, August 25, 2010, http://old.redstar.ru/2010/08/25_08 /5_01.html.
49. Aleksandr V. Khomutov, "Obshchevoiskovye formirovaniya v sovre-mennykh voennykh konfliktakh," *Arsenal Otechestva* 49, no. 5 (2020): 20.

50. Konovalov and Valetskii, *Evoliutsia chastnykh voennykh kompanii*, 105.
51. Sibilyeva, "Deyatyelnost chastnikh voenniykh kompanii v sovre-
    mennikh vooruzhenniykh konfliktakh kak viyzov mezhdunarodnomu
    gumanitarnomu pravu," , 52.
52. K. Kurilev, E. Martynenko, N. Parkhitko, and D. Stanis, "Fenomen chast-
    nykh voennykh kompanii v voenno-silovoi politike gosurdarstve v XXI
    v.," 142–143.
53. Mikhailenko set out a list of Western PMSCs believed to have been present
    in Georgia in 2008, although the list emphasised his broad definition of
    PMSCs, because it included the HALO Trust, a demining charity. Other
    private companies considered to have been active in Georgia in 2008
    include the American companies Military Professional Resources Inc.,
    Cubic Defense Applications and Kellogg, Brown and Root, as well as
    the Israeli company Defensive Shield thought to have been involved
    in training the Georgian military prior to the 2008 war. For more on
    Defensive Shield see Gili Cohen, "New Israel Police Chief's Security
    Company Trained Armies Around the World," Haaretz, August 28,
    2015, https://www.haaretz.com/.premium-gal-hirschs-security-company-
    trained-armies-around-the-world-1.5392350.
54. Sibilyeva, "Deyatyelnost chastnikh voenniykh kompanii v sovre-
    mennikh vooruzhenniykh konfliktakh kak viyzov mezhdunarodnomu
    gumanitarnomu pravu," 53.
55. Belokon, A. V. Biytyev, and L. A. Smirnova, "Chastniye voennie kom-
    panii: mirovoi istoricheskiy opiyt i sovremenniye perspektiviy dlya
    Rossii," 64.
56. Quoted in Sergey Sukhankin, "Unleashing the PMCs and irregulars in
    Ukraine: Crimea and Donbas," Jamestown Foundation, September 3, 2019,
    https://jamestown.org/program/unleashing-the-pmcs-and-irregulars-
    in-ukraine-crimea-and-donbas/. See Khomutov, "Obshchevoiskovye
    formirovaniya v sovremennykh voennykh konfliktakh."
57. See Gazprom Okhrana, https://okhrana.gazprom.ru/?. In 2007 the law
    "On the Supply of Products for Federal State Needs" stated that Transneft
    and Gazprom had the right to use weapons and security equipment to
    protect their facilities.
58. Aleksandr Mikhailenko, "Voennyie uslugi...v chastnom poryadke,"
    Krasnaya Zvezda, August 25, 2010, http://old.redstar.ru/2010/08/25_08
    /5_01.html.

59. Konovalov and Valetskii, *Evoliutsia chastnykh voennykh kompanii*, 96–99; Neelov, *Chastnyie voennyie kompanii v Rossii: opit i perspektivyi ispol'zovaniya*, 30–31.
60. Neelov, 30–31.
61. "Putin podderzhal ideyu sozdaniya v Rossii chastnykh voennykh kompanii," *RIA Novosti*, 11.4.2012, https://ria.ru/20120411/623227984.html.
62. "Pravitel'stvo RF ne podderzhalo zakonoproyekt o chastnykh voyennykh kompaniyakh," *Interfax*, March 27, 2018, https://www.interfax.ru/russia/605539.
63. The Criminal Code identifies mercenary activity as a crime, including the "recruitment, financing or other material support of a mercenary," as well as the use or participation of mercenaries in armed conflict. *Ugolovnyi kodeks Rossiiskoi Federatsii*, June 13, 1996, N 63-FZ (amended July 1, 2021, amendments entering into force August 22, 2021), http://www.consultant.ru/document/cons_doc_LAW_10699/.
64. Mikhail Gol'dreer, "ChVK net. No voprosy yest," *Arsenal Otechestva* no. 6, 38 (2018): 70–72.
65. Gol'dreer, 70–72.
66. Rostovskii, "Sergei Shoigu rasskazal kak spasal Rossiiskuyu Armiyu," 9.
67. Denis Korotkov, "'Slavyanskii korpus' vozvrashchaetsya v Siriyu," *Fontanka*, October 16, 2015, https://www.fontanka.ru/2015/10/16/118/.
68. Denis Korotkov, "Oni srazhalis za Palmiru," *Fontanka*, March 29, 2016, https://www.fontanka.ru/2016/03/28/171/.
69. Denis Korotkov, "Kukhnya chastnoi armii," *Fontanka*, June 9, 2016, https://www.fontanka.ru/2016/06/09/070/.
70. Maria Tsvetkova and Anton Zverev, "Kremlin-linked contractors help guard Venezuela's Maduro – sources," *Reuters*, January 25, 2019, https://www.reuters.com/article/us-venezuela-politics-russia-exclusive/exclusive-kremlin-linked-contractors-help-guard-venezuelas-maduro-sources-idUSKCN1PJ22M.
71. "Lavrov ob'yasnil obrashchenie Mali k rossiiskoi ChVK bor'boi s terroristami," *Interfax*, September 25, 2021, https://www.interfax.ru/world/793599.
72. "Putin otvetil na vopros o rossiiskikh naemnikakh v Livii," *RBC*, January 11, 2020, https://www.rbc.ru/politics/11/01/2020/5e19ec739a794793505 0700d.
73. Aleksei Borzhonov, "Kreml' vyiskazalsya o zaderzhaniyakh rossiyan v Belorussii: ukrainskoe grazhdanstvo ne priznayut, o ChVK

'Wagner' ne znayut," July 30, 2020, https://tjournal.ru/news/192903-kreml-vyskazalsya-o-zaderzhaniyah-rossiyan-v-belorussii-ukrainskoe-grazhdanstvo-ne-priznayut-o-chvk-vagner-ne-znayut.

74. Meia Nouwens, "China's use of private companies and other actors to secure the Belt and Road across South Asia," *Expert Commentary*, The International Institute for Strategic Studies, April 29, 2019, https://www.iiss.org/blogs/analysis/2019/04/china-bri.

CHAPTER 7

LEARNING THE WRONG LESSONS?

RUSSIA'S 2022 INVASION OF UKRAINE AND A RETURN TO "TRADITIONAL" WARFARE

Russia's failure to achieve a swift, decisive victory over Ukraine following its invasion on February 24, 2022, has raised questions about the effectiveness of its long-running military modernization and highlighted the significant gap between the theory and practice of war, particularly the dangers of ignoring fundamentals such as troop morale while being overzealous about technology. The initial Russian failures in Ukraine were all the more surprising, given the considerable financial resource and intellectual horsepower that has been invested into trying to understand how future wars might be fought and the means that an adversary may use. As discussed in previous chapters, according to the assessment of a number of Russian theorists, there had been a significant shift in armed conflict centered around how military force is exercised, with a shift from the use of massed armies to integrated, networked forces, long-range precision strike and the targeting of an adversary across all domains. Vladimir B. Zarudnitsky, head of Russia's General Staff Academy, encapsulated Russian views on the character of conflict in his 2021 analysis, stating that the development of means of warfare in the twenty-first century had stimulated a transition from the physical destruction of the adversary to a complex impact on the adversary achieved by a single integrated system that includes precision strike, reconnaissance, electronic, and information warfare that have strategic, operational, and tactical effect.[1] However, this does not reflect how the war in Ukraine has been prosecuted.

Russia's actions in Ukraine are not without precedent. Russian forces have switched to an approach that they have resorted to many times since 1991 against cities such as Grozny in Chechnya and Aleppo and Idlib in Syria: the use of heavy, indiscriminate artillery and aerial bombardments to destroy urban areas, imposing heavy costs on the Ukrainian population and their leaders. The deliberate targeting of civilians and widespread destruction of cities is intended to weaken the morale of the population in an attempt to undermine their will to resist. At the same time, the Kremlin is seeking to ensure that it is controlling information flows within Russia. It has been increasing its control over domestic media in order to ensure that the state narrative is dominant, attempting to prevent anti-war protests threatening internal stability. Gerasimov's 2018 assertion (echoing Svechin[2]) in reference to Syria that all "recent military conflicts have been significantly different from each other" is very apt.[3] The war in Ukraine is certainly a very different operation to Russia's intervention in Syria: the Russian military may have learned to fight differently in Syria, in what was characterized as an expeditionary, non-contact operation that relied heavily on the use of proxy forces, but in the process it appears to have forgotten how to fight conventional wars, particularly twentieth-century "contact" wars involving large numbers of ground forces. This chapter will attempt to draw some preliminary findings for Russian military thought based on early evidence from the war and the utility of military science's emphasis on foresight and prediction. It will also examine the extent to which lessons learned from recent operational experience have been ignored or forgotten.

In a highly critical (and prescient) article published in early February 2022, Mikhail Khodarenok, a former colonel who worked within the

General Staff's Main Operational Directorate, accused Russia's expert community of "hat-throwing fantasies" vis-à-vis a possible invasion of Ukraine. He disputed claims that Russia would be able to inflict a rapid defeat on Ukraine, arguing that "to expect to crush the armed forces of an entire nation with just one…blow is to show unbridled optimism in the planning and conduct of hostilities."[4] Warning that there was a complete ignorance of both the military-political situation in Ukraine and the level of animosity toward Moscow (described as the "most effective fuel for an armed struggle"), he cautioned that:

> No one will meet the Russian army with bread, salt and flowers in Ukraine. Events in south-eastern Ukraine in 2014 seem to have taught no one anything. Then they also expected that the entire left-bank Ukraine would turn to Novorossiya…in a matter of seconds.[5]

One of the key problems he identified with the Russian expert community's assessments of a hypothetical invasion was the assumption that Russia would have complete dominance of the air, which would facilitate a rapid victory. In his view, this ignored the lessons of Afghanistan and Chechnya, two examples of protracted wars in which the adversary did not have any air assets, as well as Georgia in 2008, where Russian forces struggled to overcome Georgian air defenses. Khodarenok's analysis provided a clear-eyed assessment of the challenges that might confront an invading Russian force, in particular the Ukrainian will to resist, and he concluded that Russia would face an uphill battle, rather than the easy one being anticipated. He also drew attention to the dangers of placing too much hope in high-tech weaponry, warning that stocks of high-precision weapons were not limitless:

> Tsirkon hypersonic missiles are not yet in service. The number of Kalibrs, Kinzhals, Kh-101 (air-launched cruise missiles) and Iskanders is at best measured in hundreds (dozens in the case of the Kinzhal). This arsenal is completely insufficient to destroy a state the size of France and a population of more than 40 million people.[6]

Khodarenok also anticipated any war in Ukraine being characterized by urban combat and warned that, because of the number of large cities in Ukraine, the Russian ground forces would likely experience a repeat of Stalingrad and Grozny, writing that "a large city is the best battlefield for the weak and less technically advanced side of an armed conflict." Despite Khodarenok's cautionary analysis, Russia's initial invasion of Ukraine appeared to be based on a number of erroneous assumptions that disregarded lessons learned during previous operations.

The continued Russian focus on time as a critical factor in armed conflict means that significant emphasis continues to be placed on decisive action during the initial phase of war, a high operational tempo, and the seizing of the strategic initiative. Although Russia initially held the strategic initiative with the decision to launch an invasion, following weeks of military build-up and speculation, it quickly lost this, largely the result of underestimating the adversary and logistical challenges, echoing the situation in Chechnya 1994, discussed in chapter 3. There was clearly a desire to achieve both military and political objectives within the shortest timeframe possible in Ukraine, which accentuated the continued focus on the initial period of war. Russia's failures to achieve a decisive victory within the first few weeks of its "special military operation" led to a protracted war of attrition, which has cost both sides dearly. Moscow was perhaps calculating that missile strikes and a multipronged invasion of ground forces would lead to a swift surrender by the Ukrainian government. Putin appears to have anticipated a repeat of Russia's decisive seizure of Crimea in 2014 or its invasion of Georgia in 2008—but the reality was more similar to its intervention in Chechnya in December 1994, when the Russian armed forces were initially unable to convert their military superiority (certainly in terms of numbers) into military and strategic success. The strength of the Ukrainian resistance appeared to surprise Moscow, prompting a change in the Russian approach as it shifted toward greater use of artillery and missile strikes against major cities, such as Kherson, Kharkiv, and Mariupol. Again, there were echoes of the Russian intervention into

Chechnya in late December 1994, when the Russian leadership planned a massive armored offensive against Grozny, intending to stage a decisive strike with air support, relying on speed to take the Chechen leadership by surprise and ensure Russia held the initiative. The Chechen forces had been long prepared for a strike against the city, and the attack was a dismal failure. The Russians underestimated Chechens' capacity to defend their homeland; similarly, the Kremlin underestimated the will of Ukrainians to defend their country.

The lessons that Russia derived from the coalition invasion of Iraq in 2003 make its initial failure in Ukraine in 2022 even more surprising. For example, a number of analysts surmised that achieving air superiority was critical in contemporary operations, something that the Russian military have never attained in Ukraine. Furthermore, Tsyganok criticized the US and its Western allies for overestimating the capabilities and impact of high-precision weapons as the decisive factor in contemporary warfare, arguing that the political isolation of leaders decides the course of war rather than military superiority.[7] Efforts to isolate Ukraine's president Volodymyr Zelensky in 2022 (and before) were ineffective, and the Russian invasion bolstered his legitimacy. There also appeared to be an (incorrect) assumption that Russia was militarily superior to its western neighbor and that this would facilitate an easy victory, with the country's leadership overlooking the fact that the Ukrainian armed forces were a very different force to that of 2014.

The war in Ukraine raises questions about Russian military science, forecasting, and the approach of its military theorists. Did they fail to heed Frunze's argument about what military science should entail, including "a correct and exact calculation of those forces and means...at the disposal of our possible opponents; and...a similar calculation of our own resources"? Has military science yet to recover from the failures of the post–Cold War era and been unsuccessful in its most fundamental task, the forecasting of future war? Have theorists become too focused on technological advances and the myth of perpetual progress, at the expense of a balanced and rounded consideration? This last question links to the central issue with Russian forecasting and military science: the focus on technology rather than intangibles such as morale, leadership, and the quality of training, which are harder for analysts to quantify but critical in battle. However, intellectual debates can only contribute to and shape political decision-making. Did political imperatives override the military reality? Certainly, the decision to invade appeared to be based on an underestimation of the adversary and an overestimation of Russian capabilities, making it likely that the realities were ignored in favor of political expediency.

FAILING TO ADVANCE?

As discussed in earlier chapters, one of the key lessons that was drawn from Western interventions was that a central objective is the weakening of a state in order to deprive it of the will to resist through both information operations and hostile encirclement. Overwhelming military superiority was also deemed to be vital, alongside network-centric warfare. According to the conclusions of a number of Russian analysts, the principal objective of warfare was no longer the destruction of an adversary's armed forces and physical seizure of territory but the ability to exhaust an adversary and undermine their will to resist. Slipchenko's concept of sixth-generation, "contactless" warfare envisaged high-precision conventional weapons and electronic warfare playing a decisive role in contemporary conflict, with ground forces in a secondary role: the principal aim is the destruction of an adversary's economic potential (including critical national infrastructure) and a change of political regime, not the seizure and holding of territory, a significant difference from previous generations of warfare. Russia's Syrian operation was held up as its first Western-style intervention fought, as much as possible, at distance, either through the use of long-range precision strike or proxy forces, a fact that prompted Defense Minister Shoigu's declaration that Russian troops had learned to fight in a new way. The use of high-precision long-range missile strikes was believed to enable a strategy of "selective action," whereby surgical strikes are directed against critical targets, inflicting high costs against an adversary at a low cost for the protagonist, who would hold the initiative.[8] Consequently, the Russian political and military leadership was perhaps overly confident in the utility of conventional high-precision weapons to achieve success during its initial invasion of Ukraine in 2022, which witnessed significant use of Russia's precision-strike capabilities such as the Iskander missile. However, during the opening days of the Russian invasion, precision strike did not have the intended effect. Moscow appeared to be working on the assumption that missile strikes and a large-scale invasion of ground forces would lead to a swift surrender by the Ukrainian government, underestimating the strength of Ukrainian resolve to resist and defend their homeland. As the war has progressed, critical national infrastructure such as power plants has been targeted by Russian forces in an attempt to undermine Ukrainian resolve and pressurize the leadership further.

This is not the first time that the Russian leadership has overlooked the role of human behavior in tipping the military balance, focusing instead on quantitative superiority, and commentators quickly began to draw parallels between Russia's performance in Ukraine and historical failures. In an article, "Unlearned Lessons from the Finnish Campaign," in *Nezavisimoe voennoe obozrenie*, Aleksandr Khramchikin, a well-known defense analyst, criticized the Russian failure to learn lessons from the 1939 Winter War between the USSR and Finland, particularly the Kremlin's underestimation of its adversary's will to resist and its apparent conviction that "Finnish workers only dreamed of joining the fraternal family of the Soviet peoples, so they would not put up any resistance."[9] The parallels between this historic campaign and Russia's failure in Ukraine in 2022 are striking: overambitious objectives set by political leaders and a Soviet force with a huge numerical superiority in terms of manpower and technology that struggled to make progress against a smaller, highly motivated adversary and suffered significant losses.

The war in Ukraine has challenged the enduring belief that technological advances constitute the central determinant of how war is fought,
while also demonstrating the criticality of intangible factors such as
morale and the will to fight and resist. War is a social construct and
human behavior remains at the heart of it. The approach of Russian
military science appears to neglect this: although there is recognition of
the importance of influence and a desire to undermine an adversary's will
to resist, much of the analysis underestimates the agency of individuals in
determining outcomes. There has also been a lot of emphasis on technological transformation: events in Ukraine have undermined much of this,
challenging assumptions about Russia's military modernization, as well
as its ability to adapt and emulate the Western experience and example.
Gerasimov's confident assertion in 2017 that Syria had proven Russian
weapons and equipment to be among the best in the world, easy to
operate and reliable, has been punctured in Ukraine. Russian weapons and
military equipment have not always performed well against an adversary
armed with advanced Western weapons such as anti-tank missiles, and
international attention has been drawn to a number of flaws in Russian-
made arms. Furthermore, Vorobyev and Kiselyev have argued that technological superiority, achieved through the acquisition of modern precision-guided weapons, had replaced mass and quantitative superiority as
the key to defeating an opponent on the battlefield.[10] Neither Russia's
precision-guided weapons nor its overwhelming numerical superiority
facilitated a swift victory, and Ukraine's supply of Western weapons
and equipment, in particular the next generation light anti-tank weapon
(NLAW), was critical in routing the Russian ground forces in the early
stages of the invasion. Russia relied upon its artillery advantage to regain
the initiative, returning to more traditional means of warfare. As the
war became more attritional, the focus shifted once more to numerical
superiority and the question of which side will have the ability to sustain
itself for the longest, in terms of weapons, ammunition, and manpower.

The high-tech warfare of the future, envisaged by a number of military
theorists, did not fully materialize in Ukraine. The war has challenged

a number of assumptions about the character of conflict in the twenty-first century, notably Slipchenko's concept of sixth-generation warfare: the war in Ukraine has been characterized by contact warfare, with adversaries facing each directly on a battlefield, rather than "contactless" warfare. Slipchenko set out three key objectives to achieve victory in sixth-generation warfare: "[the first is to] defeat the enemy's armed forces, as a rule, on his territory...[the second is to] destroy the economic the potential of the enemy, [and] the third is to overthrow or replace the political system of the enemy."[11] This analysis puts forth that comprehensive victory is no longer achieved by seizing territory but by decimating an adversary's economic potential and influencing the response of the local population to upend the existing regime. Under this framework, Russia remains a long way from achieving victory in Ukraine, where the war has been characterized by a reversion to the attritional warfare of the twentieth century, with indiscriminate artillery barrages causing widespread devastation and unifying the Ukrainian population behind Zelensky's leadership rather than undermining it. Aleksandr Raskin's 2005 observation that there is "no point in waging wars in the twenty-first century using the forms and methods from the last century and, in particular, in World War Two"[12] is pertinent in the light of the war in Ukraine, in particular the forms and methods used in both world wars. The war has borne more similarity to wars of the twentieth century than the wars envisaged by people like Zarudnitsky, Vorobyev, and Kiselyev, whose focus on technological superiority emphasized the dramatic pace of scientific and technological progress, which had led to the automation of command and control.[13] Network-centric warfare, discussed in chapter 4, was believed to constitute an evolution in warfare and the character of conflict. Although the Russian focus on NCW was driven by concern about the network-centric capabilities of its competitors (both in the West and China[14]) and fears that Russia was lagging behind in terms of its military technology, there were warnings that it was not a panacea for all the shortcomings of the Russian armed forces, clearly demonstrated in Ukraine 2022.

Russia's military, particularly the Russian Ground Forces, appeared to have forgotten the lessons learned during the years of conflict in Chechnya (and more recently in Syria), notably the importance and difficulties of urban warfare, as well as the criticality of communication and coordination between different services and units within a deployed force. Urban combat has been a central feature of Russia's involvement in Syria, something noted in the military thought literature, leading to a (perhaps misplaced) belief that important lessons had been learned, and the Russian military were well prepared. This confidence was undermined by events in Ukraine, which has been on a different scale to Russian involvement in Syria. The Russian forces sought to avoid fighting in cities such as Kyiv, which would have required a massive commitment in terms of manpower and resources such as weapons, ammunition, and equipment. One of the key lessons for the Russian military from their experience in Chechnya in the 1990s was that the use of armor in cities is risky because they are vulnerable to attack, particularly from above, from high buildings; as discussed earlier, the Chechens made good use of this approach to destroy Russian armored columns in Grozny. Hence Russian forces switched to "siege and starve" tactics against cities such as Mariupol, encircling the city, cutting it off, and then systematically bombarding it with artillery and air strikes to undermine its citizens' will to resist. This is an approach the Russians have used in the past, including in Grozny, as well as Homs and Aleppo in Syria. The deliberate targeting of civilians and extensive destruction of urban areas aims to undermine the morale of both the adversary's population and its leadership, in order to weaken their collective will to resist.

One of the key problems in Chechnya, as discussed in chapter 3, was the absence of a unified command system to manage interaction between different forces during joint operations. A similar situation took place during Russia's initial invasion of Ukraine, when there was no single operational commander in place (apart from the National Defense Management Center) until the commander of the Southern Military District (and first commander of the Russian Group of Forces in Syria)

General-Colonel Aleksandr V. Dvornikov was appointed to command the entire Ukraine campaign in April 2022, over a month after the start of military operations. Prior to Dvornikov's appointment, Russian units were competing for resources rather than coordinating their actions in pursuit of a common goal. Russian forces appeared to have problems with command and control in Ukraine, particularly vis-à-vis air-land integration, effective decision-making, and the ability to communicate over secure lines, reflecting problems experienced by the Russian military during the 2008 invasion of Georgia. The communications failings had direct consequences, with a number of senior Russian officers targeted by Ukrainian forces who had tapped their unencrypted devices, including mobile phones, to reveal their location.[15] In spite of years of investment as part of the Russian modernization program, troops on the frontline in Ukraine were often using unencrypted high-frequency radio for long-range communications and mobile phones to communicate.[16] This is despite a ban, enacted in 2020, on military personnel carrying smartphones (or other devises capable of storing photographs, videos, audio files, or geolocations) while on duty, as well as a 2019 law barring soldiers from using smartphones.[17]

The Evolving Battlefield

Nevertheless, certain aspects of Russian assessments of future war have been evident in Ukraine, notably the widespread use of precision strike (even if less effective than anticipated), along with the prominent role of both UAVs and proxy forces. Gerasimov's assessment of the characteristics of war and conflict in the twenty-first century remains fairly accurate: information superiority has been central to them, along with the increasing use of artificial intelligence, unmanned aerial vehicles, and "diverse" forces, such as private military companies.[18] UAVs have played a critical role for both Ukraine and Russia. During the early months of the war, Ukrainian forces used its Turkish-built Bayraktar TB2 UAVs to destroy a significant amount of Russian equipment in an attempt to slow Russia's advance.[19] Although their usage and success became much more limited once the Russian military focused its main effort on eastern Ukraine's Donbas—where Russia has a large number of anti-aircraft systems—the TB2 initially inflicted significant damage on Russia's military capabilities. For their part, Russian forces have used the Orlan-10 reconnaissance UAVs to support artillery fires as Iranian Shahed-136 "kamikaze" drones. Thus, the conclusions of the Russian expert community about the future of warfare were accurate about the central role of UAVs in future war. Nevertheless, despite widespread theoretical discussion and debate about their role, the Russian military appeared unprepared for the impact of UAVs during their initial invasion, highlighting the disconnect between the theory and practice of war.

Information control has been at the forefront of operations by both sides in the war in Ukraine, accentuating the enduring role of information in contemporary war and conflict. Indeed, information operations were critical to Russia's operations in Chechnya, Georgia, and Ukraine, as well as to their adversaries' maneuvers. Following Russia's invasion of Ukraine in February 2022, President Volodymyr Zelensky made appeals for Western assistance and took control of the strategic narrative, using social media to rally his people and the international community.[20] Kyiv has ensured that it is shaping the narrative, curating information from the battlefield to amplify its own messages, spotlighting Russian military incompetence. Zelensky's efforts reflected the actions of the Georgian government in 2008 under the leadership of Mikhel Saakashvili, who proved adept at strategic communications and appealing (though unsuccessfully) for Western support in the face of Russian aggression. Russia lost the battle of the narratives in 2008 and has struggled in 2022, ceding control of the information environment to Ukraine in order to focus on controlling information to the Russian population. Having learned from experience during the First Chechen War, Russian authorities have sought to rigorously control information flows, particularly on the home front, in order to shore up domestic support for military operations and to prevent these operations being undermined. In the run-up to the

2014 annexation of Crimea, access to pro-opposition sites like *Dozhd* was blocked and in 2022, the Kremlin again sought to ensure that it was controlling information flows within Russia. It further increased its control over domestic media in order to ensure that the state narrative about its "limited military operation" dominated, in an attempt to prevent anti-war protests threatening internal stability. Russia has also sought to control the information space in Ukraine's occupied areas, broadcasting Russian state television into these areas in an attempt to undermine any opposition through the use of propaganda. There have also been reports of Russian forces confiscating mobile phones and laptops.[21] Internationally, Moscow has been unable to disseminate its message to Western audiences as easily as before, because Russian media outlets such as *RT* and *Sputnik* have been shut down overseas.[22]

Russia has made very prominent use of PMSCs, such as the Wagner Group, and other proxies to bolster its regular forces in Ukraine. In March 2022, there were reports that Russia was seeking to recruit and deploy thousands of mercenaries from Syria after losing a significant number of its troops during the early weeks of the invasion. Proxy forces provided a much-needed injection of additional manpower in Ukraine: it is estimated that Russia deployed ten- to twenty-thousand mercenaries from countries such as Syria and Libya, as well as groups such as Wagner.[23] In July 2022 it was reported that Wagner operatives were being used to reinforce Russian forces on the front line and mitigate manpower issues, and "almost certainly" played a key role in the capture of the Ukrainian cities of Popasna and Lysyschansk.[24] The use of proxies represents a continuity of its approach in recent years: close cooperation with militia groups in Georgia (and Syria) gave the Russian forces a significant degree of flexibility, both operationally and diplomatically, ambiguity, and deniability. This has proved particularly useful for the Russian leadership when allegations of abuse emerge, a situation that has been repeated in Ukraine since February 2022 (and frequently since 1991).[25]

The Russian military had been learning from its operational experience in Syria, after years of observation (and emulation) of Western approaches, assessing the implications for Russia and adapting accordingly. However, its inability to achieve a quick, decisive victory over Ukraine in February 2022 demonstrated the dangers of over-reliance on technology at the expense of the basics such as troop morale, as well as the problems associated with drawing lessons from the observation of others (or your own experience). Precision strike did not have the intended effect during the opening days of the Russian invasion, raising questions about the extent to which some of the assumptions of Russian military theorists about how wars would be fought may have been flawed. Moscow appeared to be working on the assumption that long-range missile strikes and a large-scale invasion of ground forces would lead to a swift surrender by the Ukrainian government, underestimating the strength of Ukrainian resolve to defend their country. The war in Ukraine has called into question the idea that technology remains a key determinant of how war is fought. Although tangible factors such as military capabilities and technology are easy to quantify, the war in Ukraine has demonstrated the criticality of intangible factors such as morale and the will to fight and resist. Human behavior is the underpinning driver of war and conflict, increasing the inherent unpredictability and uncertainty. Attempts to impose a rigid scientific approach to understanding war may encourage misplaced confidence and the belief that all eventualities have been thought through. This overlooks the gap between theory and practice and underestimates the agency of individuals. As Khramchikin states, "when preparing an operation, the country's leadership should proceed from the real situation....[E]xercises and parades are one thing, but military operations are completely different."

Drawing lessons from the observation of others (or one's own experience) can be highly problematic. Confirmation bias may foster a focus on paradigms that support existing views or beliefs, while those that challenge or contradict these views are ignored. The interpretation of events to fit existing views is apparent in Slipchenko's analysis of the 1991 Gulf War and 1999 NATO operation: the evidence provided by these campaigns appeared to support Ogarkov's conclusions about the centrality of precision strike in contemporary warfare, amplifying Russian concerns about technological inferiority and being left behind in a new arms race. This also emphasizes the imperative of context when seeking to analyze the action of others.

Certain aspects of Russian assessments of future war have been evident in Ukraine, notably the widespread use of precision strike, along with the prominent role of both UAVs and proxy forces. Technology has played an increasingly important role in Ukraine as the war has progressed, with UAVs and electronic warfare becoming more and more significant. However, events in Ukraine in 2022 also demonstrated that technological aspects are not necessarily the only critical element of a war, the human element remains fundamental. This provides a cautionary tale about dangers of technological determinism and the myth of perpetual progress. Moving forward, the sanctions imposed on Russia as a result of its invasion of Ukraine are very likely to hinder the advance of its military, particularly the aim that technologies such as AI would play a significant role in the future development of the Russian armed forces. A lack of access to critical components, such as microchips and semiconductors, will undoubtedly have some effect on the Russian defense industrial base, particularly in the production of advanced systems and precision-guided munitions.

Notes

1. Zarudnitsky, "Kharakter i soderzhaniye voennykh konfliktov v sovremennykh usloviyakh i obozrimoi perspective," 34–44.
2. Svechin's famous quote, referenced by Gerasimov, is very applicable to
the invasion of Ukraine: "Each war represents an isolated case, requiring
an understanding of its own particular logic, its own unique character."
3. Quoted in Vladykin, "Voennaya nauka smotrit v budushchee."
4. Mikhail Khodarenok, "Prognozyi krovozhadnyikh politologov,"
Nezavisimoe voennoe obozrenie, February 3, 2022, https://nvo.ng.ru/realty/
2022-02-03/3_1175_donbass.html.
5. Khodarenok.
6. Khodarenok.
7. "Uroki i vyivodyi iz voinyiv Irake," 76.
8. Stepshin and Anikonov, "Razvitiye vooruzheniya, voennoi i spetsial'noi
tekhniki i ikh vliyaniye na kharakter budushchikh voin." They also argue
that the adoption of hypersonics means that future war will be global
in nature.
9. Aleksandr Khramchikin, "Nevyiuchennyie uroki finskoi kampanii,"
Nezavisimoe voennoe obozrenie no. 16, (April 29, 2022): https://nvo.ng.
ru/history/2022-04-28/12_1187_lessons.html.
10. They went on to argue that the growing role of IT, software modeling
and other networked activities necessitated a major change in professional military education. I. N. Vorobyev and V. A. Kiselev, 'The role of
military science in shaping the new look of the Russian Armed Forces'
Voennaya mysl' 2 (February 2011): 40–48.
11. Slipchenko, *Beskontaktnyie voinyi.*
12. Raskin, "Gryadut li 'setevyie' sechi?," 20.
13. Raskin, 20.
14. V. Kovalyev, G. Malinetskii, and Y. Matviyenko have expressed concern
about China's development of "new forms and methods of warfare,"
including psychological and information operations, as well as command and control based on network-centric capabilities. V. Kovalyev, G.
Malinetskii, and Y. Matviyenko, "Kontseptsiya 'setetsentricheskoi' voiny
dlya armii Rossii: 'mnozhitel' sily' ili mental'naya lovushka?," 95.

15. Stephen Bryen, "The fatal failure of Russia's ERA cryptophone system," *Asia Times*, May 26, 2022, https://asiatimes.com/2022/05/the-fatal-failure-of-russias-era-cryptophone-system/.

16. Sam Cranny-Evans and Thomas Withington, "Russian Comms in Ukraine: A World of Hertz," *RUSI Commentary*, March 9, 2022, https://rusi.org/explore-our-research/publications/commentary/russian-comms-ukraine-world-hertz.

17. "Putin Bans Armed Forces members from Carrying Electronic Devices, Gadgets," *Radio Free Europe/Radio Liberty*, May 7, 2020, https://www.rferl.org/a/putin-bans-armed-forces-members-from-carrying-electronic-devices-gadgets/30598888.html.

18. Gerasimov, "Razvitie voennoi strategii v sovremmenykh usloviyakh. Zadachi voennoi nauki," 10.

19. For details of the Russian military equipment reportedly destroyed by the TB2, see "Defending Ukraine - Listing Russian Military Equipment Destroyed By Bayraktar TB2s," *Oryx*, February 27, 2022, https://www.oryxspioenkop.com/2022/02/defending-ukraine-listing-russian-army.html.

20. "The role of information and social media warfare in the Ukrainian conflict," *Renforce* (Het Utrechtse Centrum voor Gedeelde Regulering en Handhaving in Europa), April 21, 2022, http://blog.renforce.eu/index.php/en/2022/04/21/the-role-of-information-and-social-media-warfare-in-the-ukrainian-conflict-2/.

21. Joshua Yaffa, "Letter from Ukraine: A Ukrainian City Under a Violent New Regime," *The New Yorker*, May 23, 2022, https://www.newyorker.com/magazine/2022/05/23/a-ukrainian-city-under-a-violent-new-regime.

22. See, for example, Council of the EU, "EU imposes sanctions on state-owned outlets RT/Russia Today and Sputnik's broadcasting in the EU," *Council of the EU*, March 2, 2022, https://www.consilium.europa.eu/en/press/press-releases/2022/03/02/eu-imposes-sanctions-on-state-owned-outlets-rt-russia-today-and-sputnik-s-broadcasting-in-the-eu/.

23. Julian Borger, "Russia deploys up to 20,000 mercenaries in battle for Ukraine's Donbas region," *The Guardian*, April 19, 2022, https://www.theguardian.com/world/2022/apr/19/russia-deployed-20000-mercenaries-ukraine-donbas-region.

24. UK Ministry of Defense (website), "Latest Defence Intelligence update on the situation in Ukraine," July 18, 2022, https://www.gov.uk/government/organisations/ministry-of-defence.

25. The Parliamentary Assembly of the Council of Europe expressed concern about what it described as credible reports of acts of ethnic cleansing committed by militia groups in South Ossetia, which Russian forces
failed to stop. Parliamentary Assembly of the Council of Europe, "The
consequences of the war between Georgia and Russia," *Resolution 1633*,
adopted on October 2, 2008. Para 13.

CONCLUSION

The systematic analysis of military trends and developments around the world, as well as the lessons learned from its own operational experience, are central to Russian military thought and part of enduring efforts to understanding the characteristics of contemporary conflict and forecast the future character of war. Forecasting the future character of war is a long-running concern, prominent in the writings of serving or retired military officers in open publications intended for an internal Russian audience. Russia's experiences of conflict during the post–Cold War era, combined with its observations of Western interventions and analysis of emerging trends and concepts, have molded military thought and views on the character of twenty-first-century conflict. The shape of future war—what it may look like and the means that an adversary may use—are a particular concern of Russian military theorists, driven partly by a recognition of historical failures (notably prior to the 1904–1905 Russo-Japanese War and World War I) to identify incremental change in warfare and the character of conflict, as well as persistent anxiety about being taken by surprise by an adversary. For states such as Russia, the lessons from Western interventions in the Federal Republic of Yugoslavia in 1999, Iraq in 2003, and Libya in 2011 have been informative, contributing to its understanding of the changing character of conflict and what this might mean for its own military.

The general view of Russia's military theorists and policy makers in the early twenty-first century was that the character of conflict was undergoing change, driven by an increase in the use of nonmilitary means to achieve strategic objectives, facilitated by changing technologies and reflecting a shift from industrial to information societies. Their conclusions were based both on Russia's own experience and observation of the operational experience of others, predominantly the US and its Western allies. Russian views of the changing character of conflict are similar to Western ones, which is not surprising as so much of their thinking has been based upon observation of Western interventions and activity. Overall, there is a widespread belief that conflict in the twenty-first century has become increasingly nonlinear and unpredictable, involving a complex array of actors, both state and non-state. The sheer pace of change, accelerated by the increased interconnectedness of societies and changes in technology and communications, has added to the complexity of the international environment, making it difficult for states to plan for future wars. The Russian focus on foresight and prediction is an attempt to ease some of the uncertainty and unpredictability that surrounds military planning.

Russian military thought continues to be heavily influenced by classical strategic thinkers such as Clausewitz, Sun Tzu, and Basil Liddell Hart, as well as Soviet military theorists who have contributed to strategic thought; Aleksandr Svechin is one of the most well-known outside of Russia for his theory of operational art. Soviet legacies continue to influence Russian military thought and the approach to understanding war and armed conflict. This legacy approach is unsurprising as many of those leading the current debates grew up in the USSR and had their thinking shaped by Soviet establishments and experiences. Furthermore, Soviet-era military theorists initiated the process of attempting to forecast the character of future war in order to understand the implications for Soviet strategy, focusing in particular on the impact of technology. In the wake of the 1991 dissolution of the USSR, Russian military science struggled to meet the needs of the state, leading to concerns that it was not addressing the kind of wars and armed conflict that Russia may face in the future, and prompting an ongoing debate about what constitutes war and conflict in the twenty-first century.

The debates among Russian military theorists about the character of conflict in the twenty-first century reveal a widely held belief that contemporary warfare has transitioned toward an indirect, asymmetric approach, based on a combination of military and nonmilitary means. Gerasimov's 2013 article, "The Value of Science in Foresight," reflected a lot of the military scientific thinking at the time, articulating a prevalent view that the character of armed conflict had undergone a qualitative change with a shift toward nonmilitary means of warfare. Armed conflict is just one of a range of instruments that a state might use in pursuit of its strategic political objectives; there is a spectrum of other means. In the opinion of many Russian theorists, actors deploy a combination of nonmilitary means, ranging from diplomatic pressure to economic coercion and information-psychological operations, underpinned by significant military capability in order to influence an adversary; the overt use of force is often only discernible later on in a conflict. An array of new instruments and enablers is envisaged, including private military and security companies, internal destabilization and "color revolutions," and networked information systems, as well as different types of media, from traditional television and radio broadcasters to social media platforms. This conviction that a wider array of actors is perceived to have the ability to exert aggressive influence and coerce an adversary also reflects concerns that the state may be losing the monopoly on the use of force.

Nevertheless, despite common themes on the character of contemporary conflict, there is debate between those Russian theorists who perceive revolutionary change occurring in warfare and the character of conflict, driven primarily by technological advances, and those who believe change to have been incremental in nature. This is exemplified by the discussions around Slipchenko's concept of sixth-generation warfare. Slipchenko discerned a fundamental shift in the character and conduct of military operations, driven by technological change (in this case the advance of precision weapons), which was disputed by those who perceive the development of precision-strike to represent only incremental change. Russia's invasion of Ukraine in 2022 challenged this focus on technological advance and superiority, validating Tsyganok's condemnation of the US and its Western allies for overestimating the capabilities and impact of high-precision weapons as the decisive factor in contemporary warfare.

There are a number of continuities in the Russian approach and the conclusions reached. Firstly, the US has long been the principal point of comparison for Russian security elites, a legacy of the Cold War and superpower competition between the USSR and the USA. Following the perceived humiliation of the 1990s, Moscow pursued parity once again and continues to compare itself with the US and be driven by concerns about its possible inferiority, both in terms of technology and global power status. Secondly, the focus on advanced technologies, in particular precision strike and networked systems, reflect Soviet-era thinking. Slipchenko's sixth-generation warfare constitutes the culmination of Ogarkov's declaration in the 1980s that an RMA was underway driven by high precision conventional weapons and information technologies. Ogarkov's assertion was made purely on a theoretical basis; observation of the 1991 Gulf War and NATO's 1999 air campaign against the Federal Republic of Yugoslavia provided Slipchenko with empirical evidence to support the theory. Continuity is also evident in the use of proxy forces, information warfare, and the central role of the cognitive domain in Russian thinking, as well as the desire to seize and maintain the initiative. Evolution can be seen in the increasing focus on PMSCs, reflecting Western practices, network-centric warfare (and electronic warfare), and the development of precision-guided munitions such as Kalibr and hypersonic weapons, as well as UAVs and autonomous systems.

Analysis of Western military experience in the post–Cold War era has allowed Russian experts to assess the relative merits of particular approaches and their implications for Russia, both in terms of its own vulnerabilities to such approaches and what it could adapt and emulate. Observation of Western interventions prompted a belief that the 1990s had heralded the development of a new paradigm of war, led by the

US and fueled by the information age: Operation Desert Storm and Operation Allied Force were deemed to be exemplars of this new approach, characterized by the central role of long-range precision strike and the critical role of information operations. One of the key lessons that was drawn from Western interventions was that a central objective is the weakening of a state in order to deprive it of the will to resist: information operations and hostile encirclement are just two ways to do this. Overwhelming military superiority was also considered to be critical, alongside network-centric warfare. Slipchenko developed the concept of sixth-generation, "contactless" warfare on the basis of these operations, asserting that high-precision conventional weapons and electronic warfare would play a key role in contemporary conflict, whereas ground forces would play a secondary one. The principal target was an adversary's economic potential, critical national infrastructure, and political regime, rather than the securing and holding of territory. There were concerns within the military theoretical community in the mid-2000s that Russia was lagging behind its strategic competitors, still focused on the contact warfare of the twentieth century.

As a result, observation has led to emulation, as seen with Russia's operation in Syria. A considerable amount of learning and adaptation has also occurred during its own operational experiences, evident in the clear transformation between the First Chechen War in 1994–1996 and ongoing operation in Syria, including the integration of automated command and control systems, and the use of UAVs, electronic warfare, and cyber operations as tactical enablers. Its own experiences appear to have reconfirmed an enduring belief in the importance of the time factor and ensuring that Russia is able to seize the strategic initiative during the initial period of war, often through the element of surprise. Modern technologies have increased the speed of events and therefore decision-making, prompting the need for improvements in the ability of the military to make decisions in order to be able to seize and maintain the initiative during the initial period of war. In Chechnya, Russian forces did not have the element of surprise and struggled to assert their authority; surprise played a key role in both Georgia and the Crimea operation. Looking toward the future, Gerasimov has emphasized a focus on preemptive actions if Russia's vital national interests are considered to be under threat. The war in Ukraine has been typified by a return to the attritional warfare of the twentieth century, with Russia returning to more traditional means of "contact" warfare and relying upon its artillery advantage to regain the initiative.

Over the first two decades of the twenty-first century, Moscow has demonstrated its readiness and willingness to use military power to achieve its strategic objectives. Russian assessments of the operational experience of both its own armed forces and of others have confirmed, on the one hand, the enduring relevance of certain key principles, including the emphasis on surprise and operational tempo, as well as the IPW and seizing and maintaining the strategic initiative. On the other hand, however, it has led to the conclusion that the role of nonmilitary means has greatly increased, particularly information-psychological means and indirect actions, and that technology has led to an evolution in warfare, with the development of long-range precision strike. Both of these changes are deemed to reduce the need to deploy hard military power to the minimum necessary. The principal objective is to exhaust an adversary and weaken their will to resist, using ambiguous, nonmilitary means to target society and social structures (rather than state structures), which are deemed to be more susceptible to manipulation. This echoes Liddell Hart's indirect approach strategy, which is intended to undermine an adversary's will to resist. This conclusion has been reached by a number of Russian analysts: the character of conflict is changing because the principal objective is no longer destroying an adversary's military and taking their territory but influencing an adversary's decision-making and transform their mindset. In the Russian view, wars are to be dominated by information and psychological warfare in order to achieve superiority in troops and weapons control, undermining an opponent's will to resist. The cognitive realm has become a battlespace, where victory is won by the domination of ideas and narratives rather than physical territory.

Information plays a central role, leading to a need to control information flows and shape the strategic narrative. The use of information as a means enables Russia to equalize what it perceives as the imbalance in the conventional military equation between itself and the US and NATO. Information operations provide a relatively low-cost, non-kinetic means of achieving its objectives, seeking to undermine an adversary's will to resist by targeting a government's ability to communicate with its population, while shaping popular opinions and exacerbating and amplifying existing societal divisions. Non-kinetic cyberattacks can be synchronized alongside kinetic activity, to complement and amplify military operations.

Thus, nonmilitary means have become a key area of focus for Russian military theorists, with particular interest in controlled chaos (color revolutions), internal destabilization, and subversion, all of which are viewed as components of Western hybrid warfare. Russia perceives a significant threat to its national security from subversion and destabilization from within. Technology is an enabler but not necessarily the critical element: what is critical is the nature of society. Societies that are open, based on liberal-democratic principles such as a free press and freedom of expression are considered to be much more vulnerable to contemporary threats, their will to resist is deemed to be much weaker. This reflects the interests of the Russian regime, the political and security elites, rather than the changing character of conflict. It is also a reflection of the country's historical experience: although fears about internal subversion and the malign use of information are not new, constituting a key part of Soviet threat perceptions, technology is. Advanced information and communications technologies enable subversive efforts, with the internet and social media platforms amplifying divisive narratives and disinformation, being tools of indirect action that target a society's ability to resist. The manipulation of interconnected, information-rich environments makes it increasingly difficult to distinguish between friend and foe.

An increasingly common thread in Russian military thought and the theoretical discourse is the perceived threat posed to Russian national identity by actors, primarily the West, seeking to undermine values and beliefs, including the belief that Russia is a great power. There is a growing focus on wars of consciousness, which are perceived to entail competition between different value systems and wherein certain systems and approaches have to be destroyed. Strategic competitors, in particular Western states such as the US, are deemed to have instrumentalized democracy promotion in order to threaten (typically illiberal) incumbent regimes and replace them with more "friendly" ones in a supposedly deliberate attempt to homogenize the world into its own image. In 2019 Gerasimov identified what he termed the Pentagon's "Trojan Horse" strategy, which he saw as intended to destabilize a country internally through the use of democracy promotion efforts. Russia's response is a strategy of active defense to preemptively neutralize perceived threats. This strategy was reportedly developed by military scientists working in conjunction with the Russian General Staff, emphasizing the real-world impact of military science. Patriotic education is also deemed to provide a vital layer of defense against the "psychological aggression" of Western adversaries, an approach that has echoes of the Soviet era, with its use of language, religion, a set of specific "national values," history, and historical memories.

In spite of this widespread belief that hostile nonmilitary measures such as information confrontation and political influence are vital components of contemporary conflict, military capabilities remain important: hostile nonmilitary means are likely to be less effective without being underpinned by significant credible military capability. The framing of internal destabilization and so-called "controlled chaos" as an external military threat to Russian national security has facilitated the expansion of defense and security responses to an essentially political issue, allowing the military to justify its involvement and interest in a range of nonmilitary activities, again in the name of "protecting" state sovereignty and regime stability.

Russian theorists detect a blurring of the boundaries between the forms and methods of waging war across all domains. A meaningful amount of Russian military theoretical literature has focused on military means and the impact of advanced technologies, what this means both for the character of conflict and, specifically, for Russia. As suggested earlier, concerns about technological inferiority, particularly vis-à-vis the US, are a leitmotif in Russian military thought. Consequently, there has been extensive observation of the operational experiences of foreign armed forces, particularly Western ones, with the apparent intention of learning lessons that could be useful for the Russian military, adapting and emulating where necessary. This observation has exacerbated Russia's enduring sense of technological inferiority, reinforcing concerns that Russia was lagging behind other states technologically, and needed to adopt concepts such as network-centric warfare and the use of precision-guided munitions, alongside the formation of integrated forces. Network-centric warfare was widely debated in Russian military journals, with specific attention paid to Western approaches and analysis of both the theoretical literature and operational experience to draw lessons for Russia. According to the assessment of a number of theorists, there has been a significant shift in armed conflict centered around how military force is exercised with a shift from the use of massed armies to integrated, networked forces, long-range precision strike, and the targeting of an adversary across all domains.

This has prompted moves to apply and integrate new technologies into the Russian armed forces, particularly information and communications technologies, to enhance the speed of decision-making and improve the efficiency and efficacy of military operations, showing a desire to take the initiative and use the element of surprise. These shifts in military thinking are reflected in the establishment of the NTsUO and the development of long-range precision strike capabilities such as the Kalibr missile. Zarudnitsky encapsulated Russian views on the character of conflict in his 2021 analysis, stating that the development of means of warfare in the twenty-first century has stimulated a transition from the physical destruction of the adversary to a complex impact on the adversary achieved by a single integrated system that includes precision strike, reconnaissance, and electronic and information warfare that have strategic, operational, and tactical effects.

As the rate of technological change accelerates and the landscape dominated by technologies that are potentially disruptive, many states envision future conflicts that involve remote and autonomous systems. Russia is no different in this respect and is already exploring artificial intelligence, weapons of new physical principles, UAVs, and other autonomous systems, evaluating the role that modern technologies can play on the battlefield. However, in a complex, interconnected world reliant on advanced technologies, the range of vulnerabilities that adversaries will seek to exploit is increasing. There is a growing recognition that in recent decades many militaries, particularly in the West, have become increasingly dependent on GPS for navigation, positioning of precision munitions and timing, to the extent that it is now recognized as a single point of failure by the US Defense Advanced Research Projects Agency. Rather than seeking to compete with multiple technologies, adversaries focus on disabling modern command, communications, and navigations systems, which would have an immediate impact on a state's military capabilities. Asymmetric operations across multiple domains offer Russia a natural counter to perceived US/Western military superiority and technological dominance.

Russian military theorists have also focused on the growing use of proxies in contemporary conflict, including PMSCs, articulating the belief that major powers have been contracting out the use of force to private companies for a range of reasons, including cost-effectiveness, reduced force numbers, a desire to avoid reputational risk, and societal aversion to casualties. The subsequent increase in Russia's use of PMSCs fits with their evolving understanding of twenty-first-century conflict and the perception that the role of non-state actors, including corporate interests, in conflict is growing. This is nothing new or groundbreaking; many scholars and analysts have explored the growing role of non-state actors in conflict in the post–Cold War era. However, it is important to view the increased prevalence of Russian PMSCs within the context of emulating Western behaviors and practices, particularly the outsourcing of risk and costs of overseas conflict to private companies. Liddell Hart's emphasis on an indirect approach, which includes physical and psychological surprise, encapsulates the Russian approach today far better than terms such as "hybrid warfare."

Drawing lessons from the observation of others and your own experience is inherently problematic. Confirmation bias can encourage a focus on evidence that supports existing views or beliefs, and disregard for evidence that challenges or contradicts existing views. The interpretation of events to fit existing views is evident in Slipchenko's analysis of the 1991 Gulf War and 1999 NATO operation: these campaigns apparently supported Ogarkov's conclusions regarding the centrality of precision-strike in contemporary warfare, amplifying Russian concerns about technological inferiority and the potential to be left behind in a new arms race. This also draws attention to the need for context when drawing conclusions from the action of others. Russian analyses of the US approach to military operations in the post–Cold War era failed to take into consideration the impact of US fears about repeating the experience of its interventions in both Vietnam and Somalia. Long-range precision strike and the perceived secondary role for ground troops in Operations Desert Storm and Allied Force were driven as much by a desire to risking US troops and avoiding casualties as by the application of a new way of war in order to gain an advantage over adversaries.

The war in Ukraine has challenged the enduring belief that technology remains the central element of how war is fought. Although tangible factors such as military capabilities and technology are easy to quantify, the war in Ukraine has demonstrated the criticality of intangible factors such as morale and the will to fight and resist. Human behavior remains fundamental to war and conflict, increasing the unpredictability and uncertainty surrounding it. Efforts to forecast the changing character of conflict and shape of future war are inherently problematic. However, this does not negate the need to understand the debates of others, particularly strategic competitors, in order to understand what they are drawing from their observations of the experience of others and themselves. While attempting to predict the future character of war and conflict, one needs to understand that adversaries are going through a similar process of observation and assessment. Russia has observed Western interventions, drawn its own conclusions, and integrated lessons from its own operational experience and culture through a process of observation, emulation, and adaptation. The lessons learned by others from one's own experiences may not be the lessons that are deemed to be the most important or comfortable, but if one fails to heed them one may be unable to effectively counter adversaries. It is thus imperative to be aware of Russian military thought, rather than merely templating Western concepts and thinking onto the actions of others. As the Russian case makes clear, states' unique experiences create different views of how military force may be used, either by themselves or potential enemies. This diversity stresses the fundamental difficulties of predicting the precise character of the next war or future conflict. Although states take different routes in attempting to manage this inherent unpredictability, they *all* seek to conduct a thorough analysis of conflicts both past and present to understand and predict how countries will fight wars in the future.

GLOSSARY

The following are terms that appear frequently in Russian military theoretical writings and have been referred to in the text.

Forms and methods of warfare (формы и способы ведения войны): forms refer to the content of military operations and include operations, engagement, combat, and strikes; methods are the procedure and techniques for the use of forces and means to solve the strategic, operational, and tactical objectives.

Disorganization (дезорганизация): disruption of an adversary's command and control/operations.

Foresight (предвидение) Seeking to account for the future development of military phenomena and processes, theory, and practice, as well as possible changes in the strategic, operational, or tactical environment. Based on intuition, common sense, and means of military experience. According to the *Military Encyclopedic Dictionary*, the capacity for foresight is an important military skill, developed during training and operations.

Great Patriotic War (Великая Отечественная война) Russian term used to describe the conflict fought from June 22, 1941 to May 9, 1945 along the Eastern Front of World War II.

Information warfare/information struggle (информационная война/информационная борьба): a frequently discussed instrument in the military's toolkit that covers both military and nonmilitary measures, depending on what is being discussed as the means or form of action. Russian thinkers view information warfare as capable of disorganizing an opponent's command and control, deceiving an adversary, sowing instability within an enemy's boxwrders, and demoralizing an opposing population or military to the point that they even lose the will to resist.

Initial period of war (начальный период войны): the most dangerous and decisive period of conflict when countries launch strategic operations with already deployed forces. According to Russian military theorists, its duration ranges from a few days to several months. May be preceded by the threatening period (угрожаемый период), of varying duration, immediately preceding the outbreak of a large-scale (regional) war when adversaries conduct military preparations. Followed by following (последующий период) and concluding periods (завершающий период).

Maskirovka (маскировка): set of military deception measures aimed at misleading the enemy about the presence, location, composition, actions, and intentions of their forces.

Military art (военное искусство): the theory and practice of preparing and conducting military (combat) operations on land, sea, and in space. An integral part of military science, which, in close cooperation with its other branches, studies the laws, forms, and methods of conducting armed struggle at the strategic, operational, and tactical level.

Military doctrine (военная доктрина): defines military-political, military-strategic, and military economic foundations for national security. Official views and positions on the character of future war and how to prepare for it.

Military science (военная наука): a system of knowledge about the current nature and laws of war, how to prepare armed forces, and modern methods for the conduct of armed struggle.

National Defense Management Center (национальный центр управления обороной Российской Федерации): designed to provide centralized combat control of the Armed Forces of the Russian Federation. Provides management of the day-to-day operations of the Air Force and Navy; collection, generalization, and analysis of information on the military-political situation in the world, on strategic directions and on the socio-political situation in the Russian Federation in peacetime and wartime. Located in Moscow.

New-generation warfare (война нового поколения): a new approach to the implementation of the concept of non-traditional warfare, incorporating both nonmilitary and military means. Often associated with the work of Sergei G. Chekinov and Sergei A. Bogdanov.

Network-centric warfare (сетецентрическая война): concept of military operations aimed at achieving information superiority through the creation of an information and communication network linking sensors (data sources), decision makers, and assets, which ensures that the participants of operations have situational awareness, accelerating command and control as well as increasing the tempo of operations.

Contactless/noncontact warfare (бесконтактная война): conflict where much of the fighting will take place via precision-guided weapons.

Nonmilitary means (невоенные действия): other levers of national power including political, information (both psychological and technical), diplomatic, economic, legal, spiritual/moral, and humanitarian measures.

Reconnaissance-fire and reconnaissance-strike complex (разведывательно-огневой и разведывательно-ударный комплекс): the coordinated use of high-precision, long-range weapons linked together with intelligence, surveillance, reconnaissance capabilities, and command and control, at the strategic level, while reconnaissance-fire is used at the tactical-operational level.

Reflexive control (рефлексивный контроль): Russian term encompassing the notion of disinformation being deployed to shape an adversary's responses and provoke a specific reaction. It gives prominence to deception and seeks to influence an adversary so that he voluntarily takes decisions that are favorable to the initiators of the action. Disinformation, deception, and *maskirovka* are key methods in the process.

Strategic deterrence (стратегическое сдерживание): a set of integrated military and nonmilitary measures to prevent and deter aggression through the promise of progressive costs and unacceptable consequences.

Strategy of active defense (стратегия активной обороны): a strategic concept integrating preemptive measures to prevent conflict and providing for measures to proactively neutralize threats to the security of the state.

Strategy of limited actions (стратегия ограниченных действий): the conduct of military operations with limited political objectives, with the deliberate spread of hostilities in strictly defined territories, using only part of the military potential or certain groups of the armed forces, selectively striking a certain number of selected objects or enemy forces.

Weapons based on new physical principles (оружие на новых физических принципах): refers broadly to weapons and military systems that operate on principles (either natural phenomena or physical processes) previously not used within the military realm such as directed-energy weapons, electromagnetic weapons, geophysical weapons, genetic weapons, and nonlethal weapons.

INDEX

Abkhazia, 81, 92–93, 196, 211
Academy of the General Staff. *See*
    General Staff
Academy of Military Science
    (AVN), 26–29, 65, 67, 164, 172,
    192, 198
active defense, strategy of, 108, 165,
    242, 250
Admiral Gorshkov, 138
Advanced Research Foundation
    (*Fond perspektivnykh
    issledovanii*, FPI), 142
Afghanistan, 69, 72–73, 151,
    201–204, 208, 212
    invasion of, 83, 85, 111, 200, 219
    lessons from, 2, 65, 75, 83, 88,
        111, 128, 199–200, 219
air defence
    S-300 and S-400 surface-to-air
        missile systems, 58
air forces (VKS), 87
air/land integration, 94
air superiority, 65, 75, 136, 221
Anapa, 143
Anikonov, Andrei, 26, 40, 47, 50,
    137, 146, 150, 232
anti-access area denial (A2AD), 40
ABM Treaty
    US withdrawal from, 64
Antiterror Orel Group, 205
Arab Spring, 156, 161–162, 179
Arbatov, Alexei, 57–58, 77, 83, 111
artificial intelligence (AI), 9, 29,
    37–38, 43, 68, 109, 123, 126, 139,
    141–144, 151–152, 173, 186, 227,

artificial intelligence (AI)
    (*continued*), 231, 244
armed forces, 5, 8, 16–17, 20, 26–29,
    33–34, 39, 41, 43–44, 53–54,
    56, 63, 65–66, 68, 70–72, 74,
    78, 81–85, 88, 91–95, 102–104,
    106–109, 111, 116, 122–123,
    125–127, 130–133, 136, 139–144,
    146, 148, 158, 171, 177, 179–181,
    191, 193–195, 197, 200–202, 204,
    207, 209, 219–222, 225, 231–233,
    240, 243, 248, 250
Armed Forces' Main Military-
    Political Directorate, 180
Armeiskii Sbornik, 6, 77–79,
    113–115, 146–149, 210
Arsenal Otechestva, 152, 206, 213
Arslanov, Khalil, 106
Arzamaskin, Yury, 178, 189
artillery, 84, 86, 88, 98–99, 111–112,
    126, 134, 137, 152, 218, 220,
    224–226, 228, 240
asymmetry, 16, 37, 71, 121, 125, 163,
    194, 237, 244
    asymmetric action, 33, 73
    asymmetric wars, 54
automation, 125, 225
Avangard hypersonic boost-glide
    vehicle, 138
Azerbaijan
    Nagorno-Karabakh, 40, 74–75,
        140, 143
    2020 Nagorno-Karabakh war, 40
    use of UAVs, 74, 109, 126, 140,
        151, 239

Babich, Vladimir, 35, 47, 49
Baluyev, Dmitry, 36
Bartosh, Aleksandr, 35, 49, 158, 160,
    179, 184–187, 189
Basayev, Shamil, 90, 196
Belarus, 178, 208
Belsky, AN, 165, 185–186
"besieged fortress," 3, 178, 180
Black Sea, 107, 143, 202
Bogdanov, Sergei, 50
bor'ba (struggle), 33–34, 48, 167
Borchev, Mikhail, 36, 49
Bordyuzha, Nikolai, 101
Borisov, Yury, 146, 166
bots, 157
Brezhnev, Leonid, 177, 184
Burenok, Vasily, 142, 151–152
Bush, George W., 161

Caspian Sea, 104
Caucasus, 9, 81, 87, 90, 94, 104, 113,
    195–196, 199, 203
    mountains, 92
Central African Republic, 205, 207
Central Military District, 105–106
Center for Military-Strategic
    Studies (TsVI), 27
Center for Reconciliation of
    Opposing Sides (Syria), 199
Chechen battalions (in Ukraine)
    Dzhokhar Dudayev
        peacekeeping battalion, 199
    Sheikh Mansur peacekeeping
        battalion, 199
"Chechenization," 91, 197
Chechnya, 196
    Interim Council, 91, 197
    Russian invasion of, 7–8, 54, 77,
        81–89, 91–92, 94–100, 104,
        107–109, 111–113, 126, 128,

Chechnya
    Russian invasion of (continued),
        197, 200, 212, 218–221, 226,
        228, 239
    siege of Grozny, 85–86, 91,
        97, 111–112, 126, 197, 218,
        220–221, 226
Chekinov, Sergei, 24, 26, 28, 37–38,
    46–47, 50, 54, 56, 70, 72, 77, 79,
    126, 129, 136, 146, 148, 150, 159,
    173, 184, 187, 195, 208, 211, 249
Chernysh, Anatoly, 133, 149
Chief of the General Staff. See
    General Staff
    See also Gerasimov, Valery, and
        Ogarkov, Nikolai
China, 26, 74, 128, 131, 141, 148,
    216, 225, 232, 265–266
    information warfare, 170–171,
        187
    use of private military and
        security companies (PMSC),
        202–203, 209
Clausewitz, Carl von, 32, 236
    definition of war, 36
    warfare, 2
Clinton, Bill, 161
coercion, 113, 175, 237
    coercive use of force, 58
cognitive operations, 174–175
Cold War, 4, 6, 8–9, 24–25, 40, 46,
    54–55, 59, 69, 71, 76, 82, 144,
    160–162, 171, 177, 187, 191–192,
    196, 199–201, 208–209, 211, 221,
    235, 238, 245
Collective Security Treaty
    Organization (CSTO), 101
color revolution, 2, 9, 37, 155–156,
    161–165, 237, 241
Combat Control Center, 132

Combat Control Center (*continued*)
  See also National Defense
    Management Center
    (NTsUO)
command and control, 19, 28,
    40, 43, 65, 67, 69, 71, 84, 87,
    101–103, 105–106, 108, 122, 125,
    132, 144, 148, 172–174, 227, 232,
    247, 249
  automated command and
    control, 56, 123–124, 131,
    133, 135, 137, 142, 239
  See also network-centric warfare
    (NCW), 5, 127–131, 133–134,
    147, 167, 225
C4ISR (command, control,
    communications, computers,
    intelligence, surveillance, and
    reconnaissance), 121, 128, 167
communications, 54, 61, 68, 70, 74,
    87–88, 94–95, 97, 102, 106, 134,
    161, 167, 175, 227–228, 265
  capabilities, 67, 101, 104–105,
    121, 123, 128–129, 139, 144,
    243–244
  technologies, 9, 121, 123, 127,
    132, 139–140, 144, 157,
    235–236, 241, 243–244
Confederation of Mountain Peoples
    of the Caucasus (KNK), 196
conscient war, 186
conscription, 205
controlled chaos, 9, 155–157, 160,
    163–164, 183, 241–242
Cossacks, 92, 192, 198
counterinsurgency operations, 67,
    83, 87, 91, 111, 197, 265
Crimea, Russian annexation of,
    98–99, 101, 179, 229
cyber, 34, 96, 98, 114–115, 125, 142,
    156, 169, 173, 182, 187, 203, 239

Danilenko, Ignat, 25, 45, 47
deception. See *maskirovka*
defense, 29, 43, 53, 56–57, 61–64,
    67, 75, 85–86, 94–95, 100, 102,
    104–106, 108, 111, 124, 126, 131,
    134, 136, 138, 142, 163, 165–166,
    173, 178–181, 183, 185–186,
    200–201, 214, 222–223, 226, 231,
    242, 244, 248, 250, 265
  collective, 58
  Ministry of Defense, 26–27, 44,
    46, 48, 87, 116, 122, 132, 139,
    143, 198, 205–206, 233
democracy, 160, 162, 185
  promotion of, 73, 161, 183, 242
destabilization, internal, 155–157,
    161, 182–183, 237, 241–242
distributed denial-of-service
    (DDoS) attacks, 96
  Georgia 2008, 98
deterrence, 212, 266
  nuclear, 63, 136–137
  strategic, 63–64, 137, 142, 250
dialectical materialism, 21
Directorate for Innovative
    Development, 143
disinformation, 90, 157, 159, 167,
    169, 173–175, 241, 249
disorganization, 247
  of forces, 134, 136, 174–175
domains, 40, 203, 243–244
  air, 61, 68, 106, 182, 238
  information, 61, 66, 166, 177,
    180, 182, 217–218, 238
Donskov, Yury, 134, 149–150, 173,
    188
drones. See UAVs
Duma, 85, 162–163, 179, 185,
    205–206
Dvornikov, Aleksandr, 39, 50,

Dvornikov, Aleksandr (*continued*), 103–104, 106, 115–116, 227
Dzhokhar Dudayev, 84, 91, 197
  peacekeeping battalion, 199

electronic warfare (EW), 29, 57, 59, 61, 66–67, 72, 74, 98, 101, 103, 106–107, 109, 123, 129, 131, 133–135, 139–140, 143–144, 147, 167, 169, 171, 173, 175, 182, 222, 231, 238–239
  electronic shock, 56
ERA Technopolis, 143
exercises. *See* military exercises

Federal Security Service (FSB), 87, 91, 162, 206, 211
fifth columnists, 165, 178
Finland
  1939 Winter War, 223
forecasting, 2–3, 5, 7, 15–16, 23–28, 42, 44–47, 58, 65, 77, 126, 141, 221–222, 235
Foreign Policy Concept (Russia), 97
foresight, 2–3, 5, 7, 10, 15, 17–18, 21–22, 25, 28, 42, 44–45, 126, 164, 218, 236–237, 247
Frunze, Mikhail, 20–21, 221
futurology, 25

Gantemirov, Beslan, 91, 197
Gareev, Makhmut, 28, 46, 77
Gazprom, 205, 214
General Communications Directorate, 106
General Staff, 27, 73, 165, 178, 217, 219, 242
  Academy of the General Staff, 24, 58, 107–108, 132
  Chief of the General Staff, 4, 28, 168

Georgia, 7–8, 54, 72, 81–82, 91, 95, 97–98, 101, 114, 122, 128, 196, 211, 214, 219–220, 227, 229, 240
  Abkhazia, 92, 197
  National Security Council, 93
  opolchentsy, 92, 197–198
  president of, 96, 99, 228
  Roki Tunnel, 92
  South Ossetia, 9, 92–94, 111, 212, 234
  Tskhinvali, 92, 203
  2003 Rose Revolution, 161
Gerasimov, Valery, 1, 19, 24, 28, 33–34, 36, 38–39, 42, 44, 46–51, 73, 79, 102, 105, 108, 115, 117, 122, 131–133, 139, 146–147, 165–166, 178, 186, 192, 199, 210, 218, 224, 227, 232–233, 237, 240, 242
  speeches by, 98, 104, 109, 164, 185
  writings of, 15–16, 18, 29, 164
Gorbunov, Viktor, 36–37, 49–50, 68–69, 77–78, 156, 184
Gracheva, Tatyana, 194, 211
Great Patriotic War, 18, 29, 32, 62, 85–86, 247
grey-zone operations, 1
Gromyko, Yu, 166, 172, 186
Ground Forces, 56–57, 59, 67, 73, 86, 88, 93–94, 102, 107, 124, 134, 137, 140, 145, 218, 220, 222–224, 226, 230, 239
Grozny, 91, 97, 111–112, 126, 197, 218, 221
  New Year''s Eve offensive, 85
  siege of, 226
  urban combat, 85–86, 220, 226
Gulf War, 1991, 4, 8, 24–25, 53, 56–59, 65, 72, 135, 231, 238, 245

Gulf War, 1991 (*continued*)
  See also Operation Desert Storm

Hezbollah, 108, 199
highways. *See* infrastructure
Hmeimim (Russian base), 105, 199
hybrid warfare, 1, 98, 121, 158, 184,
  192–193, 241, 245
humanitarian intervention, 58
humanitarian operations, 104
hypersonic missiles, 122–123, 125,
  131, 136, 138, 143, 150–151, 219,
  232, 238
  See also Avangard, Tsirkon

Independent International Fact-
  Finding Mission on the Conflict
  in Georgia, 114, 212
indirect action, 156, 159, 182, 195,
  241
influence operations, 162
information, 34, 78, 88, 93, 155, 163,
  195, 212, 229, 236, 266
  confrontation, 5, 29, 35, 39, 42,
    56, 61, 70, 96–97, 101, 103,
    138, 142, 145, 156, 158–159,
    164, 166–170, 175–177,
    181–182, 242
  domain, 66, 166, 177, 180, 182,
    238
  narratives, 96, 166, 228, 240–241
  operations, 4–5, 54, 56, 61,
    65–68, 72, 84, 89, 95–96, 98,
    100–101, 103–105, 108–110,
    115, 123, 125, 127–130,
    132, 134, 138, 144, 148, 156,
    158–159, 164, 169–171,
    173–175, 177–178, 181–182,
    222, 228, 232, 237, 239, 241,
    243, 248–249

information (*continued*)
  superiority, 37–38, 55–56, 61,
    66–68, 70–71, 83–84, 108,
    110, 124–126, 129–130, 134,
    144, 168, 171–175, 222, 227,
    239–240, 249
  systems, 9, 27, 29, 56–57, 61,
    65–68, 103, 105, 108–109, 122,
    124–127, 129–130, 132–134,
    137, 142, 144–145, 160,
    166–168, 170–171, 174–176,
    182, 185, 227–228, 237–238
  warfare, 4–5, 9, 27, 29, 38, 42–43,
    46, 54–57, 59, 61, 66–67,
    71, 89–91, 95–98, 100–101,
    103, 108–110, 123, 125–127,
    129–130, 134, 138, 144,
    147–148, 156–158, 167–182,
    186–187, 218, 222, 227–228,
    232–233, 237–241, 243–244,
    247, 249
information-psychological
  confrontation, 36–37, 39, 71,
    158, 168, 170, 175–179, 195, 237,
    240
information-technical
  confrontation, 168, 170, 175
information technology (IT), 1–7,
  9–10, 16–24, 26–28, 32–34,
  36–37, 41–43, 45, 47, 51, 53,
  56–59, 61–62, 64, 67–71, 81–82,
  85–86, 89–90, 92–98, 100–104,
  106–109, 113, 121–126, 128–133,
  136, 138–139, 143–147, 155–158,
  160–163, 165–175, 180, 182–183,
  192, 194, 196–198, 201, 203–209,
  212, 214, 218, 220, 222, 224–226,
  228–229, 232, 234–236, 238–241,
  244–246, 248–249
infrastructure, 86
  communications, 61, 94, 134

infrastructure (*continued*)
  critical national, 94, 168,
    222–223, 239
  economic, 57, 61, 71, 222, 239
  highways, 94
  military, 55, 57–58, 61, 71,
    93–94, 109, 134, 136, 168, 205,
    222–223, 239
  pipelines, 94
  railway, 94
  targeting, 55, 71–72
  transport, 61, 94
initial period of war (IPW), 19–20,
  45, 84, 93, 109, 130, 134, 167, 220,
  239–240, 248
initiative, 18, 20, 69, 71, 83, 85,
  93–94, 99–100, 107, 109, 124,
  130, 136–137, 144, 158, 167, 195,
  206, 209, 221, 223–224, 238–239,
  243
  maintaining, 19, 240
  seizing, 19, 84, 220, 240
insurgency, 44, 100, 111
integrated forces, 40, 108, 129, 243
integrated operations, 107, 140
integration, 27
  air/land, 94
  forces, 87, 91, 94, 103, 110, 123,
    127–129, 131–132, 139, 227,
    239
intelligence, surveillance, and
  reconnaissance capabilities
  (ISR), 137
internal instability, 2, 5, 43, 157,
  162–163
international law, 37, 58, 181
invasions, 18–19, 22, 32, 66, 78,
  83–85, 88, 90, 111, 182
  of Georgia, 7–8, 54, 72, 81–82,
    91–99, 101, 114, 122, 128, 161,
    196–197, 203, 211–212, 214,

invasions
  of Georgia (*continued*), 219–220,
    227–229, 234, 240
  Mongol, 3
  Napoleon, 3
  of Russia by Germany (1941), 3
  of Ukraine, 1, 6–10, 40, 54,
    74–75, 81–82, 97–102, 105,
    109, 115, 121, 123–126, 137,
    140, 143–145, 161, 178, 192,
    198–200, 203, 206–208, 212,
    214, 217–221, 223–233, 237,
    240, 245
Ionov, MD, 175, 188
Iraq, 40, 74, 128, 151, 163, 201–205,
  208, 212
  invasion (1991 and 2003), 2, 8,
    25, 53, 61, 64–69, 72–73, 75,
    78, 131, 135, 221, 235
  Mosul, 67, 78
  weapons of mass destruction, 66
Iran, 74
irregular forces, 100, 192–193
ISIS, 107
Iskander missile, 145, 219, 223
Isserson, Georgii, 18, 20, 22

Kadyrov, Akhmad, 91, 197
Kalibr missiles, 104, 123, 138–139,
  219, 238, 243
Kalinovsky, Oleg, 113, 171–172, 187
Kalistratov, Aleksandr, 193, 210
Kartapolov, Andrei, 156, 159, 180,
  184
Kavkazcentre, 90
Kazar'yan, Bogdan, 129–130, 148
Kenya, 205
Kepel, Oleg, 178, 189
Khamzatov, Musa, 3, 11, 23, 36, 46,
  49, 77, 125, 130, 146, 148, 193,
  210

Khodarenok, Mikhail, 218–220, 232
Khomutov, Aleksandr, 50, 88, 113, 149, 203, 213–214
Khramchikin, Aleksandr, 148, 223, 230, 232
Kiselyev, Valery, 47, 67, 78, 85, 112–113, 125, 128–129, 135–137, 146, 150, 159, 194–195, 208, 210, 224–225
Komov, Sergei, 171–172, 187
Kondratyev, Georgy, 86, 111
Konovalov, Ivan, 194, 203, 211, 214–215
Korobov, Igor, 107
Kosovo, 57, 77, 83, 111
Kostenko, Alexei, 67, 78, 210
Krasnaya Zvezda, 47, 50, 90, 116–117, 148–152, 184, 186, 201–202, 213–214
Kruglov, Vyacheslav V., 20, 25, 38, 45–47, 49–50, 74, 79, 150, 187
kto kogo, 19, 32
Kudryavtsev, Aleksandr, 193, 210
Kulikov, Anatoly, 127, 147
Kurochko, Mikhail, 35, 49, 156, 184
Kutishchev, Viktor, 94, 114, 147

Lapin, Aleksandr, 105–106
lasers, 122, 141
Lefebvre, Vladimir, 175
Lenin, Vladimir, 32, 48
Liddell Hart, Basil, 33, 156, 158, 184, 211, 236, 240, 245
Libya, 2, 8, 40, 53, 72–74, 192, 204, 207–208, 229, 235
limited actions, strategy of, 108, 250
Limno, Aleksandr, 170, 187–188
logistics, 67, 132, 193, 210
Lomaia, Alexander, 93
Lutovinov, Vladimir, 37, 50, 180, 189

Main Operational Directorate, 219
Makarov, Nikolai, 163, 204
Makhonin, Viktor, 5, 11, 35, 45, 47, 49, 77
Mali, 207, 215
Mann, Steven, 160–161, 185
Mariupol, 220, 226
Marxism-Leninism, 18, 32
    dialectical materialism, 21
maskirovka (military deception), 169, 174–175, 249
Messner, Evgenii, 157
Middle East and North Africa, 163
Migunov, Aleksandr, 170, 186–187, 189
Mikhailenko, Aleksandr, 202–205, 213–214
Mikryukov, Vasily, 176, 189, 192, 202, 210, 212–213
military art, 4, 21–23, 67, 103, 124, 170, 248
military exercises, 100, 107, 140, 175
    Kavkaz, 106
    Vostok-2018, 106, 116
    Zapad-2021, 143, 153
military intelligence, 207
military reform, 95
military science, 10, 22, 24, 30, 39, 42, 58, 102–103, 109, 126, 132–133, 135, 146, 148, 171, 201, 218, 221–222, 224, 232, 248
    Academy of Military Science (AVN), 26–29, 65, 67, 164, 172, 192, 198
    definition of, 16–17, 28, 36, 45
    Soviet, 3, 7, 15–16, 20–21, 23, 25–26, 31, 45–47, 77, 198, 236, 242
military theorists, 2, 4–6, 16, 19–20,

military theorists (*continued*), 22, 29, 33, 37, 40, 43, 53–55, 62, 67, 75, 90, 95, 124, 134, 140–141, 170, 178, 193–194, 221, 224, 230, 235–237, 241, 244, 248
military thought, 1–2, 4–7, 9–11, 15–17, 20, 22–23, 28, 32, 42, 53, 75–76, 95, 105, 122–123, 140, 144–145, 155, 176, 182, 187–188, 192, 195, 201, 209, 218, 226, 235–236, 242–243, 246
military technological revolution (MTR)
 See also revolution in military affairs (RMA), 4, 55, 124, 238
militia, 212, 234
 Abkhazian, 92, 197
 South Ossetian, 92, 197–198
 use of, 104, 108–109, 197–199, 229
 in Ukraine, 198–199, 229
Ministry of Defense, 26–27, 44, 46, 48, 87, 116, 122, 132, 139, 143, 198, 205–206, 233
Ministry of Emergency Situations
 Ministry of Emergency Situations' Crisis Center, 132
Ministry of Foreign Affairs, 115, 206
Ministry of Internal Affairs (MVD), 132, 198
Ministry for the Press, Television and Radio Broadcasting, 89
missile defense, 63, 124, 136, 138
 Czech Republic, 64
 interceptors in Poland, 64
 See also ABM Treaty
Mizintsev, Mikhail, 132
modernization
 of armed forces, 8, 54, 81, 122–123, 128, 134, 217, 224,

modernization
 of armed forces (*continued*), 227
Moiseev, Vladimir, 38, 50
Molkino (Wagner training base), 207
morale, 29, 48, 124, 145, 159, 204, 217–218, 222, 224, 226, 230, 245
Moran Security Group, 205
Mozambique, 207
Munaev, Isa, 199
multi-domain operations (MDO), 40, 106

Nagorno-Karabakh, 143
 2020 Autumn War, 40, 74–75, 140
nanotechnology. *See* technology
narratives
 information, 96–97, 101, 166, 180, 218, 228–229, 240–241
 strategic, 96–97, 101, 166, 228, 240–241
National Bank of Georgia, cyber attacks against, 96
National Defense Management Center (NTsUO), 43, 100, 226, 248
NATO (North Atlantic Treaty Organization), 5, 51, 56, 61–62, 68–69, 73–75, 77, 79, 82–83, 87–88, 115, 122, 128, 131, 164, 169, 186, 231, 238, 241, 245
 collective defense, 58
 Operation Allied Force, 8, 53–54, 57, 59, 84, 135, 239
Neelov, Vladimir, 194, 200, 211, 213, 215
Nerekhta, 143, 153
new generation war, 71
new physical principles, weapons of, 37, 43, 140, 244

network-centric warfare (NCW),
    5, 9, 40, 43, 55, 67, 71, 95, 123,
    127–131, 133–134, 144, 147, 167,
    222, 225, 238–239, 243, 249
networked systems, 40, 127,
    130–131, 133, 238
next generation light anti-tank
    weapon (NLAW), 224
Nezavisimoye voennoye obozrenie
    (NVO), 6, 151–152
Nigeria, 205
non-governmental organizations,
    165
non-kinetic action, 34–35, 41–42,
    54, 72, 95–96, 156, 169–170, 191,
    212, 235–236, 241, 249
non-military means, 91, 95
non-state actors, 72, 82, 191–194,
    208, 244–245
North Caucasus, 8–9, 81, 87, 90, 92,
    104, 195–196
Nogovitsin, Anatoly, 168
nuclear weapons, 4, 11, 55, 62, 64,
    68, 75, 124–125
    deterrence, 63, 136–137

Ogarkov, Nikolai, 4, 55
Operation Desert Storm, 8, 53–55,
    75, 239
Orange Revolution, 2004, 161
Orlan-10 reconnaissance UAV, 228
Orlyansky, Vladimir, 124, 146,
    172–173, 187–188

Pankov, Nikolai, 94
patriotic education, 179, 242
    military-patriotic education, 180
Patrushev, Nikolai, 162
Persian Gulf, 63
pipelines. See infrastructure
Popov, Igor M., 23, 36, 46, 49, 77,

Popov, Igor M. (continued), 193, 210
Popov, Vladimir V., 133, 149
Potekhin, V, 166, 172, 186
Prague Spring, 1968, 158
Pravyi Sektor (Right Sector), 198
precision, 5, 43, 64, 75, 102, 106,
    122, 224, 249
    strike, 4, 9, 40, 55, 58, 63, 73,
        81, 104–105, 123–124, 126,
        134–138, 140, 144–145,
        165, 217–218, 222–223, 227,
        230–231, 238–240, 243–245
    guided munitions, 38, 56, 60, 62,
        72, 131, 135, 138, 231, 238,
        243
    weapons, 37, 59, 61–62, 65,
        86, 95, 104–105, 125, 127,
        135–136, 145, 223, 237–238
prediction, 6, 10, 17–18, 28, 68, 218,
    236
preemptive action, 19, 38, 125, 165,
    168, 240, 242, 250
private military and security
    company (PMSC), 9, 38, 102,
    108, 110, 143, 191, 193–195, 198,
    200, 204, 210–211, 227, 237–238,
    244–245
    Article 13 of the constitution,
        206
    Belarus, 208
    Blackwater, 201–203, 208, 213
    Chinese use of, 202–203,
        208–209, 216
    Defensive Shield, 203, 214
    draft legislation, 205–206
    Greystone, 203
    Military Professional Resources
        Inc., 203, 214
    presence in Africa, 192, 203,
        205–207
    SBS Othago, 203

private military and security
company (PMSC) (*continued*)
Prigozhin, Yevgeny, 207
Venezuela, 207, 215
Wagner Group, 10, 192, 203,
205–208, 216, 229
professional military education,
108, 146, 232
Prompt Global Strike, 138
propaganda, 33, 66, 83, 89–90, 167,
169, 173, 229
protest potential, 42, 158, 164–165,
178, 182
proxy forces, 9, 81, 83–84, 91–92,
98, 100, 178, 196–197, 218, 222,
227, 231
private military and security
company (PMSC), 102, 108,
110, 192–195, 198–202,
208–209, 229, 238
psychological effects
See also psychological means,
instruments
psychological warfare, 71, 110,
176–180, 189, 240
See also psychological
operations
Putin, Vladimir, 8, 31, 115, 185

railways. *See* infrastructure
reconnaissance, 27, 40, 67, 69–70,
72, 105–106, 108, 121–122, 124,
129, 131, 135–136, 140, 143–144,
147, 218, 228, 244
reconnaissance fire, 137
reconnaissance-strike, 104, 107,
137–139, 249
reflexive control, 171, 174–176, 188,
249
regime change, 43, 157, 194

regime change (*continued*)
US, 2, 5, 37, 161–162
Iraq, 2
Hussein, Saddam, 63, 66
Reznichenko, Vasily, 62, 78
revolution, 144, 157
color, 2, 9, 37, 155–156, 161–165,
237, 241
in military affairs, 3–4, 11, 46, 59
military technical, 4, 21
robotic systems, 102, 109, 122–123,
126
Nerekhta, 143, 153
Uran-9, 143, 152–153
Rodionov, Sergei, 72, 79, 171, 187
Rog, Valentin, 63, 78
Rogov, Sergei, 64, 78
Roki Tunnel, 92
Rosatom, 132
Rose Revolution (2003), 161
Rosgvardiya (National Guard), 206
Roskomnadzor, 101
Rossiiskaya Gazeta, 90, 185
Runov, Valentin, 72, 79
Russian invasion of
Chechnya, 7–8, 54, 81–85,
87–89, 91, 94–97, 99–100,
107, 109, 113, 126, 128, 197,
200, 212, 218–221, 226, 228,
239
Georgia, 7–8, 54, 72, 81–82, 91,
93, 95–99, 101, 114, 122, 128,
196–197, 203, 212, 219–220,
226–229, 234, 240
Ukraine, 1, 6–9, 54, 81–82,
97–102, 105, 109, 115, 121,
123–126, 137, 139–140,
143–145, 178, 192, 197–200,
203, 207–208, 217–221,
223–231, 233, 237, 240,
245–246

Russia
    as a besieged fortress, 3, 178, 180
    competitors, 62, 131, 155, 160,
        162, 178, 225, 238–239, 242,
        246
    public opinion of military
        operations, 88, 96, 167–168,
        176
    threat perception, 41, 58
Russian Federation, 90, 163, 185,
    249
    armed forces, 5, 8, 15–17, 20,
        26–29, 33, 39, 41–44, 53–54,
        65, 67–68, 70, 72, 74, 81–85,
        87–88, 91–95, 99, 102, 104,
        106–109, 111, 116, 122–133,
        135–136, 139–144, 146,
        148, 158, 171, 174–181, 191,
        193–194, 196–197, 200, 202,
        204, 207, 209, 217, 219–222,
        224–225, 231–232, 240–241,
        243, 248
    constitution, 205–206
    Military Doctrine, 26, 36, 41, 194
    military modernization, 8, 54,
        81, 122–123, 128, 134, 217,
        224, 226–227
    Ministry of Defense, 26–27, 44,
        46, 48, 87, 116, 122, 132, 139,
        143, 205–206
    Ministry of Internal Affairs, 132
    National Security Strategy, 41
    president of, 48, 51, 115, 190,
        211, 213
    Russian State Armament Program,
        123

Saakashvili, Mikhel, 96, 228
Saifetdinov, Kharis, 16, 44, 126, 133,
    146–147, 149
sanctions, 123, 145, 163, 175, 231,

sanctions (*continued*), 233
satellites, 105–106
satellite communications, 106
Serbia, 2, 8, 57–59, 61, 88
    2000 Bulldozer revolution, 161
Serdyukhov, Anatoly, 95
Shahed-136 UAV, 228
Sharp, Gene, 162, 185
Sheikh Mansur peacekeeping
    battalion, 199
Shoigu, Sergei, 115–116, 132, 135,
    201, 206, 213, 215
Shubin, Aleksei S., 20, 38, 45, 50, 74,
    79, 150–151
sixth-generation warfare. *See*
    Slipchenko, Vladimir
Slavonic Corps, 206
Slipchenko, Vladimir, 24, 31, 46, 57,
    60, 62, 64, 67, 72–74, 77–79, 112,
    125, 134–135, 144, 149, 222, 230,
    232, 239, 245
    generations of war, 59
    sixth-generation warfare, 8, 54,
        58–59, 61, 63, 65, 71, 75, 157,
        225, 237–238
Smolvy, Aleksandr, 27, 73, 79
Snesarev, Andrei, 166
social media, 157, 176, 228, 233, 237,
    241
societal divisions, 145, 241
    inciting, 156
soft power, 9, 35, 155
Sokolovsky, Vasily, 4, 11, 17, 33, 41,
    44–45, 48, 51
South Ossetia, 9, 93–94, 212, 234
    Joint Peacekeeping Forces, 92
    Tskhinvali, 92, 203
Southern Military District, 103, 226
sovereignty, state, 58, 183, 242
space, 9, 27, 29, 39–40, 42, 61, 96,
    101, 103, 122, 125–126, 129, 133,

space (*continued*), 135, 159,
    161–162, 165–166, 169, 181–182,
    192, 196, 208, 229, 248
    space-based assets, 66–67, 105,
        138
    space reconnaissance systems,
        124
special forces, 104, 109, 136, 158,
    164, 207
Stalingrad, 85–86, 220
Stepshin, Mikhail, 26, 40, 47, 50,
    137, 146, 150, 232
strategy, 11, 17, 22, 26–27, 29, 32,
    41, 44–45, 47–48, 51, 56, 58,
    91, 121, 136–137, 155, 159–162,
    184–185, 188–189, 192–193,
    195, 197, 211–212, 223, 236, 240,
    265–266
    active defense, 108, 165, 242, 250
    limited actions, 108, 250
strategic parity, 55, 64, 75, 123, 136,
    138
Strategic Nuclear Forces Control
    Centre
    See also National Defense
        Management Center
        (NTsUO)
strategic vulnerability, 3
subversion, 2, 33, 37, 39, 43–44, 56,
    100, 136, 156–158, 177, 241
Sun Tzu, 158, 236
superiority
    military, 18–20, 37–38, 40, 56,
        61, 64–66, 71, 75, 94, 105,
        108, 110, 124–126, 129–130,
        134–138, 144, 168, 170–171,
        173, 175, 192–193, 202,
        220–225, 227, 237–240, 244,
        249
    information, 37–38, 55–56,
        61, 66–68, 70–71, 83–84,

superiority
    information (*continued*), 108,
        110, 124–126, 129–130,
        134–135, 144, 168–175, 222,
        227, 239–240, 249
surprise, 7, 18–19, 65, 84–85, 93–94,
    99–100, 107, 109, 111, 124, 131,
    135, 144, 167, 175, 195, 220–221,
    235, 239–240, 243, 245
    fear of, 3, 8
Suvorov, Aleksandr, 93, 168
Svechin, Aleksandr, 20, 22, 32,
    44–45, 48, 69, 79, 218, 232, 236
swarm tactics, 107
Syria, 1, 7–10, 39–40, 73–75, 81–82,
    87, 99, 101–104, 106–109, 116,
    123–124, 132, 135, 137–140, 143,
    149, 151, 176, 192, 200, 206–207,
    224, 229–230, 239
    Aleppo, 126, 218, 226
    Center for Reconciliation of
        Opposing Sides, 199
    Idlib, 126, 218
    Hmeimim, 105, 199
    Homs, 226
S-70 Okhotnik UAV, 140
S-300, S-400 surface-to-air missile
    systems, 58

Tagirov, Viktor, 105
technology, 2, 4, 7, 11, 26, 39, 48, 56,
    65, 70, 74–75, 90, 106, 122, 124,
    126–127, 131, 139–140, 142, 145,
    151, 163, 167, 169, 217, 222–223,
    225, 230–231, 236, 238, 240–241,
    245
    emerging, 9, 22, 55, 143–144,
        147, 192–193, 235, 265
    nanotechnology, 37, 43, 68, 125,
        143

technological inferiority, sense of, 3, 55, 243
technological enablers, 102
terrorism, 158, 207
    Chechen, 81–86, 88–91, 95, 111–113, 196–197, 199, 211, 221, 228, 239
    international, 19, 27, 34–35, 37, 53, 57–58, 64, 66, 68, 75, 89, 92, 96–99, 101, 107, 111, 114–115, 128, 141, 160, 163–164, 181, 183, 185, 188–189, 193, 197, 201–202, 206, 211–212, 216, 224, 228, 236, 265
    terrorist groups, 72
threat, 2, 5, 9, 26, 32, 37, 42–43, 62, 64, 66, 98, 100, 140, 155–157, 162–165, 169, 173, 175, 179, 181, 183, 186, 195, 240–242
    perception of, 41, 58
Tikhanychev, Oleg, 137, 150
Transnistria, 9, 196
Triandafillov, Vladimir, 20, 22, 45
troll factories, 157
Tskhinvali, 92, 203
Tsirkon hypersonic missile, 138
Tsyganok, Anatoly, 65, 102, 115, 221, 238
Tukhachevskii, Mikhail, 20
Tulip Revolution, 2005, 161
tunnels
    tunnel warfare, 107
Turkey, 209
    Libya, 74
    support for Azerbaijan, 74
    Syria, 74
    TB2 Bayraktar UAV, 75, 140, 227–228, 233
    UAVs, 74
Tyutyunnikov, Nikolai, 129, 148,

Tyutyunnikov, Nikolai (continued), 211

Udugov, Movladi, 90
Ukraine, 1, 6–10, 40, 54, 74–75, 81–82, 97, 99, 101–102, 105, 109, 115, 121, 123–126, 137, 140, 143–145, 192, 198, 200, 203, 206–208, 212, 214, 217–219, 221, 223, 225, 227, 230–233, 237, 240, 245
    Dzhokhar Dudayev peacekeeping battalion, 199
    Kyiv, 98, 100, 199, 226, 228
    Lysyschansk, 229
    Maidan, 178
    Mariupol, 220, 226
    next generation light anti-tank weapon (NLAW), 224
    Orange Revolution, 161
    Popasna, 229
    Sheikh Mansur peacekeeping battalion, 199
Unified Information Space, 133
United Nations Security Council (UNSC), 58
United States, 2, 11, 15, 26, 37, 46, 54, 59, 61, 65–70, 73–74, 76, 78, 96–97, 107, 121–122, 127–128, 130–131, 141, 151, 155, 162, 164–165, 169–170, 173, 177, 182–183, 185, 196, 198, 201, 207–208, 215, 221, 236, 239, 241, 243–245, 266
    competition with, 238, 242, 265
    fear of, 56
    missile defense, 57–58, 63–64, 136, 138, 145
    National Endowment for Democracy, 160

United States (*continued*)
1994 National Security Strategy, 161
nuclear weapons, 4–5, 55, 62–64, 75, 136, 142–143
strategic parity, 55, 64, 75, 136, 138, 145
withdrawal from ABM Treaty, 64
unmanned aerial vehicles (UAVs), 139, 142
Orlan-10, 228
S-70 Okhotnik, 140
Shahed-136, 228
TB2 Bayraktar, 75, 140, 227–228, 233
unmanned underwater vehicles, UUVs, 139
Uran-9, 143, 152–153
urban combat, 84–87, 107, 112, 126, 220, 226
USSR, 55, 161, 177
dissolution of, 236
ideological struggle, 32
Red Army, 19, 180
World War I, 18, 20
Utkin, Dmitry, 207

Valdai discussion club, 31
Valetskii, Oleg, 194, 203, 211, 214–215
Venezuela, 207, 215
Vestnik Akademii voennyikh nauk, 6, 47, 50, 79
Vinogradov, Vladimir, 135, 150
Vladimirov, Aleksandr, 33, 45, 48, 202, 213
Voennaya mysl,' 5–6, 11, 24, 26, 44–47, 49–51, 74, 77–79, 112–114, 117, 142, 146–152, 179, 184–189, 210–211, 213, 232

Voenno-promyishlenniyi kur'er, 5, 37, 77, 185
voina, 5, 9, 11, 33–34, 46, 49, 77, 89, 115, 167, 210
Vostok-2018, 106, 116
See also military exercises
Vostok battalion, 198, 212

Wagner Group, 216
in Africa, 192, 205, 207
Belarus, 208
Prigozhin, Yevgeny, 207
training base, 207
in Ukraine, 9–10, 192, 203, 206–208, 229
Venezuela, 207
See also PMSCs and Utkin, Dmitry
war, 6, 10, 19, 29, 32, 44–45, 47, 55, 63–66, 69, 79, 81–88, 91–94, 96, 98–99, 101, 109, 111–114, 127–128, 132, 140, 142–143, 145, 150, 157–163, 172, 177–180, 185–186, 196–199, 201, 203, 209, 211–212, 214, 218, 220–221, 224, 227, 229–231, 234, 240, 245, 266
bor'ba, 33–34, 48, 167
character of, 1–5, 7–8, 15–17, 20–22, 24–27, 31, 33–34, 36, 39–41, 43, 53–54, 56–58, 62, 67–68, 73–76, 125, 135–136, 139, 141, 144, 164–167, 181, 191–194, 200, 202, 204, 208, 217, 225, 232, 235–236, 243, 246, 248
classical, 23, 34–35, 56, 126, 156, 170, 236
debate around, 1–2, 4, 15–18, 23–24, 28, 31, 33–34, 36, 56, 135–137, 167, 171, 205–206,

war
  debate around (*continued*), 228,
    236
  new generation, 37, 54, 56,
    59, 61–62, 70–72, 75, 102,
    125–126, 129–130, 134, 144,
    187, 222–223, 238–239
  non-classical, 35, 156
  opening and closing phase, 38,
    72
  World War I, 20, 22, 235
  World War II, 18, 247
  voina, 5, 9, 11, 33–34, 46, 49, 77,
    89, 115, 167, 210
warfare, 2, 4, 18, 20, 26–28, 37–38,
    41–42, 68, 83, 85, 88–91, 94,
    96–97, 100, 110–111, 122, 156,
    160–161, 168, 170, 172, 174,
    176–181, 186–189, 191, 200, 204,
    213, 217, 221, 224, 226, 228, 233,
    235, 247, 265
  contact, 46, 54, 57, 59, 61–63,
    125, 135–136, 225, 239–240
  electronic, 29, 56–57, 59, 61,
    66–67, 72, 74, 98, 101, 103,
    106–107, 109, 123, 129–131,
    133–135, 138–140, 144, 147,
    167, 169, 171, 173, 175, 182,
    218, 222, 231, 238–239, 244
  generations of, 59–60, 222
  hybrid, 1, 98, 121, 158, 184,
    192–193, 241, 245
  network-centric, 5, 9, 40,
    43, 54–55, 67, 71, 95, 123,
    125–128, 130–131, 133,
    144, 147–148, 222, 225, 232,
    238–239, 243, 249
  non-contact/contactless, 46,
    54, 57, 59, 61–63, 65, 125,
    134–136, 222, 225, 239, 249
  sixth generation, 8, 46, 54,

warfare
  sixth generation (*continued*),
    58–59, 61, 63, 65, 71, 75,
    125, 134–135, 157, 222, 225,
    237–239
Warsaw Pact, 158
weapons, 26–29, 31, 35, 38–40,
    63, 65–66, 71, 73–74, 76, 86,
    95, 102–105, 108, 110, 122–123,
    126–130, 133–134, 137, 139,
    141–142, 152, 159, 167, 171–172,
    175–176, 191, 214, 219, 221, 224,
    226, 237, 249–250
  conventional, 4, 55–56, 59, 61,
    75, 81–82, 109, 121, 125,
    135–136, 145, 222–223,
    238–241
  of new physical principles, 37,
    43, 140, 244
  nuclear, 4, 11, 19, 55, 59, 61–62,
    64, 68, 70, 75, 124–125,
    131–132, 136, 143
West, the, 2, 5, 7, 9–10, 39, 43, 46,
    53, 57, 62, 64, 67–68, 72–74,
    96–97, 99, 102, 105, 111, 122,
    124, 126, 128–129, 131, 147, 155,
    158, 160, 163, 167, 169, 177–180,
    183, 192, 196, 201–204, 207–209,
    214, 221, 225, 228–230, 236,
    238–239, 241–242, 244–245
  concepts, 6, 8, 40, 54–55, 75, 95,
    121, 123, 127, 235, 243, 246
  doctrine, 1, 41, 194
  intervention, 8, 58, 65, 75,
    81–84, 98, 106–107, 109, 162,
    197–199, 222
  next generation light anti-tank
    weapon (NLAW), 224
Western interventions, 6, 39, 54–55,
    68, 95, 222, 236, 238–239, 246
  Afghanistan, 2, 65, 69, 72–73, 75,

Western interventions
  Afghanistan (*continued*), 83–85,
    87–88, 200–201, 204, 208–209
  Iraq, 2, 8, 53, 64–66, 69, 72–73,
    75, 201, 204, 208–209, 235
  Kosovo, 83–84
  Libya, 2, 8, 53, 72–74, 235
Western Military District, 105, 107,
  180
will to resist, 37, 65, 70, 87, 96, 100,
  111, 145, 167, 176–177, 218–219,
  222–224, 226, 239, 241, 248
  undermining, 25, 33, 38–39, 58,
    64, 66, 69, 72, 83–84, 95, 99,
    110, 130, 155–158, 168, 178,
    182, 225, 240
Winter War, 1939, 223

Yegorov, Nikolai, 85
Yugoslavia, Federal Republic of, 57,
  235, 238
  NATO intervention, 58
  See also Kosovo, Operation
    Allied Force
Yunarmiya, 180, 190

Zapad-2021, 143, 153
  See also military exercises
Zarudnitsky, Vladimir, 20, 45, 50,
  73, 79, 136, 141, 150, 152, 167,
  178, 186, 189, 194, 208, 211, 217,
  225, 232, 243
Zelensky, Volodymyr, 221, 225, 228
zero-sum logic, 19, 32
Zhuravlev, Aleksandr, 107, 140
Zolotov, Leonid, 88, 113

Cambria Rapid Communications in Conflict and Security Series

General Editor: Thomas G. Mahnken
(Founding Editor: Geoffrey R. H. Burn)

The aim of this series is to provide policy makers, practitioners, analysts, and academics with in-depth analysis of fast-moving topics that require urgent yet informed debate. Since its launch in October 2015, the RCCS series has the following book publications:

- *A New Strategy for Complex Warfare: Combined Effects in East Asia* by Thomas A. Drohan
- *US National Security: New Threats, Old Realities* by Paul R. Viotti
- *Security Forces in African States: Cases and Assessment* edited by Paul Shemella and Nicholas Tomb
- *Trust and Distrust in Sino-American Relations: Challenge and Opportunity* by Steve Chan
- *The Gathering Pacific Storm: Emerging US-China Strategic Competition in Defense Technological and Industrial Development* edited by Tai Ming Cheung and Thomas G. Mahnken
- *Military Strategy for the 21st Century: People, Connectivity, and Competition* by Charles Cleveland, Benjamin Jensen, Susan Bryant, and Arnel David
- *Ensuring National Government Stability After US Counterinsurgency Operations: The Critical Measure of Success* by Dallas E. Shaw Jr.
- *Reassessing U.S. Nuclear Strategy* by David W. Kearn, Jr.
- *Deglobalization and International Security* by T. X. Hammes
- *American Foreign Policy and National Security* by Paul R. Viotti

- *Make America First Again: Grand Strategy Analysis and the Trump Administration* by Jacob Shively

- *Learning from Russia's Recent Wars: Why, Where, and When Russia Might Strike Next* by Neal G. Jesse

- *Restoring Thucydides: Testing Familiar Lessons and Deriving New Ones* by Andrew R. Novo and Jay M. Parker

- *Net Assessment and Military Strategy: Retrospective and Prospective Essays* edited by Thomas G. Mahnken, with an introduction by Andrew W. Marshall

- *Deterrence by Denial: Theory and Practice* edited by Alex S. Wilner and Andreas Wenger

- *Negotiating the New START Treaty* by Rose Gottemoeller

- *Party, Politics, and the Post-9/11 Army* by Heidi A. Urben

- *Resourcing the National Security Enterprise: Connecting the Ends and Means of US National Security* edited by Susan Bryant and Mark Troutman

- *Subcontinent Adrift: Strategic Futures of South Asia* by Feroz Hassan Khan

- *The Next Major War: Can the US and its Allies Win Against China?* by Ross Babbage

- *Warrior Diplomats: Civil Affairs Forces on the Front Lines* edited by Arnel David, Sean Acosta, and Nicholas Krohley

- *Russia and the Changing Character of Conflict* by Tracey German

For more information, see **cambriapress.com**.

www.ingramcontent.com/pod-product-compliance
Lightning Source LLC
Chambersburg PA
CBHW050341270326
41926CB00016B/3565